The Black Glass Encyclopedia

The West Virginia Museum of American Glass, Ltd.

The book committee:
Neila Bredehoft
Helen Jones
Tom Felt
Jaime Robinson
Dean Six
and
Bob Jones, lead photographer

Schiffer Publishing Ltd®

4880 Lower Valley Road, Atglen, PA 19310 USA

This book is dedicated to the glass collecting passion of Doris and Roy White.

Library of Congress Cataloging-in-Publication Data

The black glass encyclopedia / The West Virginia Museum of American Glass, Ltd.
p. cm.
Includes bibliographical references and index.
ISBN 0-7643-2312-1 (hardcover)
1. Black glass—United States—Catalogs. I. West Virginia Museum of American Glass.

NK5439.B54B53 2005
748.2'0973'03—dc22
2005017605

Designed by "Sue"
Cover design by Bruce Waters
Type set in Bernhard Modern BT/Souvenir Lt BT

ISBN: 0-7643-2312-1
Printed in China
1 2 3 4

Published by Schiffer Publishing Ltd.
4880 Lower Valley Road
Atglen, PA 19310
Phone: (610) 593-1777; Fax: (610) 593-2002
E-mail: Info@schifferbooks.com

This book may be purchased from the publisher.
Include $3.95 for shipping.
Please try your bookstore first.
You may write for a free catalog.

In Europe, Schiffer books are distributed by
Bushwood Books
6 Marksbury Ave.
Kew Gardens
Surrey TW9 4JF England
Phone: 44 (0) 20 8392-8585; Fax: 44 (0) 20 8392-9876
E-mail: info@bushwoodbooks.co.uk
Free postage in the U.K., Europe; air mail at cost.

Contents

Thanks and Acknowledgments

Special Thanks

Much of, indeed a large percentage of, the photography in this volume is the work of Bob Jones. For his efforts, time, and expertise we owe a huge debt. Thanks and well done Bob!

The WEST VIRGINIA MUSEUM OF AMERICAN GLASS (WVMAG) is, at its best, a group of friends who love glass. We love its history, the objects, learning about it, and sharing it. We are also dedicated to preserving the legacy of the American Glass Industry: from archival paper to photographs and images, from glass objects to tools and endless interesting parts of the story!

It should come as no surprise then that WVMAG's mission, indeed our passions, lead us to take our collections and knowledge to you, the curious readers, wherever you may be. We would like to extend a sincere invitation to you to visit the physical Museum, located in Weston, West Virginia. However, for many reasons you might not be able to come to us. Thus we hope our growing projects of books, such as this volume on black glass, and our wildly popular series of Monographs and catalog reprints will bring some of our work and passion to you. Please also consider visiting us on line at www.allaboutglass.org or at http://members.aol.com/wvmuseumofglass/. One further way to share our journey in learning about glass is to become a museum member. Membership earns you the acclaimed quarterly color magazine *All About Glass*.

This book you hold, like all things accomplished by WVMAG, is the work of a number of fun, dedicated and anxious to share people. On behalf of the book committee for this volume we thank all of you who have supported us and shared with us to make this book yet another of the Museum's dreams come true!

Those who shared images, beyond our principal photographer Bob Jones, are these kind folks who have earned our sincere thanks:

JoAnne & Earl Autenreith
Larry Baker
Tom Bredehoft
Debbie & Randy Coe
Tom Felt
Steve Ockner
Kelly O'Kane
Dick Krause
Dale Murschell
David Schepps

Those who shared their glass to be photographed, thus meriting our thanks, include:

Joyce and Jim Coverston
John & Susie Determan
Janet Gould
Sally Lockard and David Bush
Bob Page and Dale Frederiksen
Gary Geiselman and Allyn Rosa
Nelson Linger
C.J. Mack
George Muraskevics
Replacements, LTD
Noel Pletcher
Susie Thayer
Donna and Bob Marks
Tiffin Glass Collectors Club
Roy & Doris White
Loren Weyant

Beverly Kappenman merits our thanks for allowing us use of some of the extensive information from her well organized, useful web site on black glass (http://members.aol.com/blkapp). We cannot say enough positive about this marvelous resource and tool for exploring black glass. Many thanks to Beverly from all of us!

Tim Schmidt was kind in sharing his knowledge of Central Glass. His continued support of WVMAG on many levels is appreciated.

Roger and Claudia Hardy, Akro Agate Glass authorities, whose willingness to share and support glass projects is always appreciated.

Thanks goes to The Tiffin Glass Collectors Club and museum for their sharing copies of catalog pages used in the company history chapter. The spirit of cooperation between museums is always a breath of fresh air and we all win!

Some pieces in the book are from the collection of The WEST VIRGINIA MUSEUM OF AMERICAN GLASS, Weston, West Virginia.

Randy and Debbie Coe continue to dream and share a vision for a West Coast glass museum. Their enthusiasm and passion for glass excites us here in Weston, West Virginia. Their real gift to WVMAG is their ability to get excited about our vision for our museum and the work that comes with that. The Coe's shared with us several dozen excellent photographs that appear in this book. For their support of this and other projects we affirm our thanks and friendship.

Several hundred of the pieces of black glass photographed for this book were from the collection of Doris and Roy White, long time members of the West Virginia Museum of American Glass and active in its founding. Shortly after the three-day photography session for this book was held, much of the White's collection was sold at public auction. Some of the pieces sold now reside at WVMAG. Many people purchased pieces knowing those objects would appear in this book and they are indeed in here! The opportunity to include the White collection was a major factor in the success of this project. Doris and Roy merit special thanks. This book, like many Museum projects, could not have been completed without their dedication and willingness to share. Doris and Roy, we hope this is like a scrap book for you and helps you to continue to "hold" and enjoy your collection for many years to come. Many of us can now share in enjoying your collection with you!

An Introduction to Black Glass

How This Book Works

This book has two parts; for all practical purposes there are two books in this binding! One book shows photographs of objects arranged alphabetically by shape, for which you should see the table of contents. There are captions with relevant data such as dimensions, dates, manufacturers (when known), and pricing information. The second "book" is about who made the black glass. This section is arranged alphabetically by the name of the glass company and includes period advertisements and catalog pages. The history section shows many objects we did not have access to for individual photographs. This allows attributions of hundreds of additional black glass pieces. Some pieces may be illustrated in both sections, many will appear in only one or the other. You are urged to explore both parts of the book when seeking to identify a black glass object.

Also please be alert to the possibility that an object may be represented in the book by a different form in the same pattern. Thus, if you have a sugar bowl and cannot find your object identified as a sugar bowl, it may be shown as a candlestick or whatever other object was made available for photography. You might find the same pattern identified but illustrated by a different shape. We did not exclude multiple forms of one pattern and have included as many as we had access to. Do please remain open to this as a possible avenue to identifying an object.

Please note we had difficulty in arranging some objects within a specific chapter. What appeared a covered candy dish one day seemed to us to be a covered compote the next day. When original manufacturers names are known, we have followed those. When no such designation was found, we did our best. And we did not all agree at all times on what a piece was or where it best fit. Thus, we urge you to explore similar chapters to find a piece if it does not readily present itself in the chapter you first explore.

Finally, do understand that hundreds and hundreds of shapes of black glass have been produced in the US and around the world that we have not included in this book. We photographed hundreds of objects for inclusion. And we know there is sufficient black glass out there to create an entire series of similar black glass books, which, at this time, we do not aspire to do. So, accept that many of the most common forms are represented in this work as are some uncommon and a few truly rare objects. Should your black glass piece not be found, do not give up. See the bibliography at the end of the book and, using similar objects found in this volume, continue your quest with our best wishes for success.

If after cruising this book your identification questions about stemware or tableware persist and you have no answer, another good source for answers is www.replacements.com. An amazing number of images of black and black in combination objects are illustrated on the site, but it may take some seeking as patterns are arranged by manufacturer's name and not by color – but browsing can be entertaining!

Should the glass object you seek to identify not be found in the shape chapters of this book remember to look in the period illustrations in the company history chapters. It may be there!

A final note on using this book: we have sought to confirm for you our attribution as to manufacturer and other relevant data. We have often found reason to question why some authors make attributions. For that reason we have sought to make our attribution calls as transparent as possible for you. We tell you, in parenthesis, what our sources are for most attributions. The parenthetical references all are detailed in the book's bibliography. When multiple sources aided in identification we have cited the earliest of the source that showed us catalog pages or other primary source material over those simply stating an alluded attribution. This serves two purposes: you can explore our reasoning and draw your own conclusions if in doubt and it gives a wonderful jumping off point if you wish to seek further information. We hope this part of our work provides answers where you might have questions.

This book itself is a part of The WEST VIRGINIA MUSEUM OF AMERICAN GLASS Living Books Project. Glass scholar, author, and researcher William Heacock set a wonderfully high standard for all of us when, in his books and glass magazines, he consistently printed updates and corrections of information found in his diverse earlier works. He was not afraid to admit he had made mistakes. He welcomed new, supplemental, additional, and emerging data as we researched and discovered it. The telling of history is not a stagnant thing. As we learn, the accounts and what we tell must change. Thus, this book, like many other antiques and collectibles books, must be open to new discoveries that change what we believe and put in print.

New black glass information will emerge, in part prompted by the publication of this book, and in part as new interests arise and discoveries of additional catalogs and archival materials are shared. And this is the operative word, "Shared."

The West Virginia Museum of American Glass launches, with the publication of this book, an addition to the *All About Glass* web site called The Living Books Project. The goal is to create a site where new material that compliments, supplements, and corrects the hard print book can be posted and shared. It is NOT about updating prices but supplementing information. New prices will remain the turf of future books and authors.

Time will tell if this concept is workable, practical, and used. We hope it is. We invite you to go to www.allaboutglass.org and follow our Living Books Project there.

The Pricing in This Book

Most books on collectibles today include, by necessity, pricing information attempting to aid the collector. Some books have undertaken this task by stating the value they believe an object to be worth. This practice attempts to "set" the market price. Other books, such as this one, attempt to reflect the current price on the market, thus reflecting the current reality or "trend." Considerable effort has been taken to explore and document prices given in this volume. However, the numbers shown for values are only an attempt to determine current market values and are not advice or absolute determinations. They should be used as a guide, suggesting a range of value. This is information you might take into consideration when determining what price you would pay for or sell an object for. Note that you must, absolutely must, also consider other factors.

Some of the factors to be considered in establishing an item's price include condition, the interest and supply/demand of similar objects in your immediate market, and an awareness of national trends. Electronic internet auctions offer instant and nationwide access to an amazing variety of objects with real and current sales. The immediacy of such electronic auctions reflects values and prices realized that can and do shift quickly and frequently. Price ranges offered in this book are suggestions, taking into consideration the information we could gather and should be used only as suggestions from which you should determine your own understanding of value or price.

Chapter One
Animals and Figurals

Butterfly #5629 BK. Fenton Art Glass. 3 1/2 inches tall. Fenton Gift shop item, not in cataloged line. 2003. $25-30.

Squirrel on base #670. Dalzell-Viking Glass. 5 1/2 inches tall. Acid mark on base "Dalzell." Made sometime between 1987-1998. $40-55.

Rearing Horse #6200 Bookends. L.E. Smith Glass. Each, $80-95.

Swan Bowl. Golden Ray line with brass metal bases attached. Unattributed as to the glass producer but probably New Martinsville Glass. The name Golden Ray was registered to G.H. French Co. This style was popular in the 1950s. See the Kappenman black glass web site where this is illustrated and explained. Web site noted in Introduction chapter. $18-26.

Swan centerpiece bowl #922/IS. Viking Glass. 10 inches. Clear applied neck. Appears in 1953 Viking price list. *WVMAG Permanent Collection*. $65-85.

Swan open dish. Boyd Glass from Cambridge Glass Co. mold, hand painted. Marked "B" in diamond on bottom. Circa mid-1980s. (Coe 2003) $24-28.

Swan #3/10. L. E. Smith Glass. 5 3/4 inches tall x 8 3/4 inches. See image following for mark on base. Circa 1920-30s. (Weatherman 1974) $34-46.

Mark appearing on bottom of open bowl L.E. Smith swan. Mark reads "Hand-Made Glassware L.E. Smith" in oval and "MADE IN USA" beneath oval. The same swan bowl is found more often without the mark.

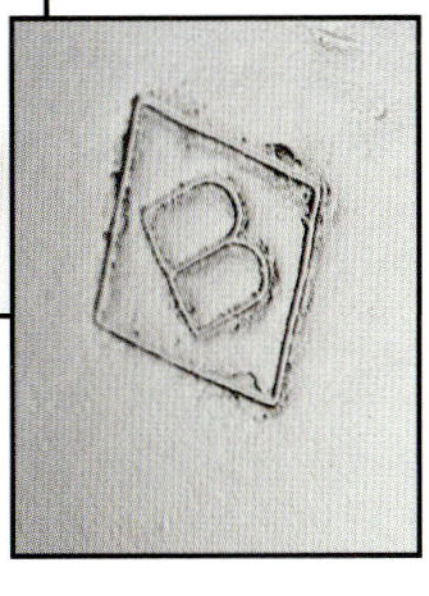

Detail of "B" in diamond mark found on bottom of black swan from Cambridge mold. Boyd Glass Co. mark used circa mid-1980s.

Swan open bowl. Cambridge Glass. 4 1/2 inches in length. Marked "C" in a triangle on base. $34-42.

Swan #974 open dish. Viking Glass. 4 3/4 inches. Circa 1950s. $30-38.

Swan open bowl #1324. Viking Glass. 12 inches in length. Identical swans in smaller sizes and a variety of colors were made. Circa 1950s. $45-65.

Swan, handmade in the offhand whimsey style with crystal tail and wings. $10-18.

Kingfisher. Imperial Glass manufactured from original Heisey mold. Never made in black by Heisey. 4 inches tall. Original has holes in base for use as flower frog, removed when reissued by Imperial. Circa 1982. $150-225.

Hen on Nest, miniature. Degenhart Glass. 3 1/2 inches, two part. Marked with a "D" in the outline of a heart on base, typical Degenhart mark. $12-18.

Duck on Nest. Unknown manufacturer. Two part. 7 inches in length. Marked "S". Found in cobalt and other colors, believed to be contemporary and it has been suggested it is Smith Glass but we have not confirmed this. $24-38.

Bird. Scandia Glass. 4 inches tall. Handmade and part of an extensive off-hand line of glass animals made by Scandia and many of its nearby neighbor factories in West Virginia. Distinguishing these handmade animals is nearly impossible without labels or strong provenance. *Part of the collection at WVMAG from the Scandia factory.* $8-12.

Chanticleer #2629. Fostoria Glass. Produced 1953-1957. 10 1/4 inches tall. Also produced with gold decoration. $300-350.

Happiness Bird #5197. Fenton Art Glass. Copper Roses decoration on black. Produced circa 1989-1993. (Walk 2004) $35-45.

Scottie Dog #6653. L.E. Smith Glass. 5 inches tall. Circa 1930s. (Coe 2003) $40-60.

Dog covered dish #570. Co-operative Flint Glass Co. Shown without lid. Produced circa 1927. With lid. $350-425.

Bull Dogs. 2 1/2 inches tall. Note that several glass houses manufactured these; the color and detail suggest these are Westmoreland Glass "Black Mist." $40-48.

Scottie Dog Bookend #1288. Imperial Glass from Cambridge Glass mold circa 1979. Satin 6 1/2 inches tall. $125-155.

Wolfhound #716. Dalzell-Viking Glass. 7 inches tall. Found marked "Dalzell" with acid stamp. Originally produced in crystal only by New Martinsville Glass pre-1944, and continued in the 1940s by Viking. Most, if not all, black pieces will be from the Dalzell-Viking era 1987-1998. $80-95.

Epic Cat #1322. Dalzell-Viking Glass. 8 1/2 inches tall. Originally Viking Glass, part of the Epic line. Most, if not all, black solid figural pieces will be from the Dalzell-Viking era 1987-1998. (Six 2003) $35-50.

Sassy Susie and Grotesque Cats. U.S. Glass. Left: Sassy Susie Cat #9448, satin and hand painted, 11 inches tall, 1924-1936. $225; Right: Grotesque cat #9446, hand painted, 6 1/2 inches tall, 1924-1934. $175

Tiger Paperweight #103. Imperial Glass Corporation. Made in 1982. Limited production of 705. Marked "ALIG" reissued from original Heisey mold. 8 inches long. (Coe 2003) $150-180.

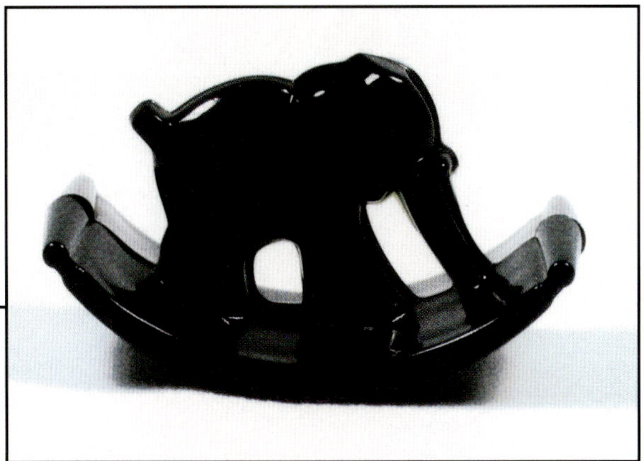

Elephant ink rocker blotter. Houze Glass. Circa 1930s. Note that four forms were made in this ink rocker or blotter series and each is found in a different color: elephant = black; Donkey (colt) = opaque yellow; Scotty dog = jade; sailboat = cobalt blue. As shown, $45-65.

Elephant covered dishes. Co-operative Flint Glass. Smaller $250-350; larger $550-700.

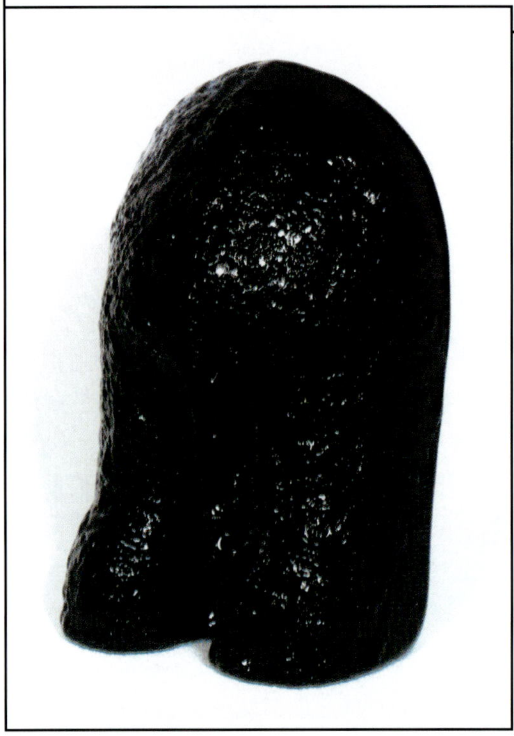

Elephant. Pilgrim Glass. 5 inches tall. Handmade at some time between 1987-2001. $15-20.

Elephant bookend #9531. Blenko Glass. 6 inches tall. Note that a "proper set" includes both ends of the animal. As shown, $10-18.

Polar bear. Pilgrim Glass for Oneida. Mystic Animal Collection. A series of native animals as talisman. The bear was to be marketed with two small glass "stones" on his back and a piece of rawhide to tie them together then place the piece inside a leather pouch. Pilgrim was incapable of producing the quantity needed or of delivering the speed Oneida needed, so the series prototypes were sold in the Pilgrim factory outlet. With original box, etc., $40-60. Bear alone, $10-15.

Whale. Scandia Glass. 5 1/2 inches, length may vary as these are handmade. The nose was where the piece was held when hot to work the glass so some "pontil" like scars may exist if not well fire polished. Several neighboring glass companies made identical and difficult to distinguish animals including Whales. *Collection of WVMAG from the Scandia factory.* $8-14.

Santa in Sleigh covered dish #1972-1. Westmoreland Glass. 5 1/2 inches in length. (Coe 2003) $40-55.

Buddah Chinese bookend #2626. Fostoria Glass. Circa 1957. 7 1/4 inches tall. $350-500.

Chinese Lute #2626. 12 inches. Fostoria Glass. Gold decoration #522. Produced 1954-1957. $350-500.

Chinese Lotus #26261/2. Fostoria Glass. 12 inches. Gold decoration #522. Produced 1954-1957. $350-500.

Venus Rising figural. Imperial Glass. Original Cambridge Glass Co. mold for flower frog reissued, but without flower holes for Mirror Images in 1981. 7 inches tall. (Coe 2003) $40-50.

Mushroom. Viking Glass. One of four sizes produced. This is the large and is 3 inches tall. The jumbo is 4 inches tall and 6 inches across. Two smaller sizes were made. See original Viking catalog pages in the company history chapter. (Six 2003) $22-32.

Cathay Bookends. Imperial Glass Corporation. 7 inches. Made in 1982. Marked "ALIG." On left is #5030 Lu-Tung book holder and on the right is #5029 Empress. These were reissues of the original pattern made by Imperial in the 1949-1953 era but originally only in crystal. $200-275 each.

Chapter Two
Bed and Bath Items

Atomizer Perfume #7948. Fenton Art Glass. Hand painted enamel flower design. 4 1/2 inches tall, circa 1990. $32-48.

Perfume with stopper. Satin. Not American manufacturer. 6 inches tall. Believed to be modern – post 1980. $8-10.

Perfume with stopper. Unknown manufacturer. Modern – post 1980. 5 1/4 inches tall. $8-18.

Perfume with crystal stopper. Acid mark "Made in Czechoslovakia" with paper label reading "Genuine Marlee Glass." Circa 1930s. 4 3/4 inches tall. $120-200.

Crystal bottle with black applied threading and black stopper. Steuben Glass. 5 1/2 inches tall, square. $225-250.

Left:
Bristol Yellow bottle #6887. Steuben Glass. Black applied threading and black stopper. Square. $275-350.

Right:
Perfume bottle with crystal stopper. Believed to be contemporary. Unattributed. $18-26.

Three toed powder box #12 with lid. Duncan & Miller Glass. 4 1/2 inches diameter. Gold decoration. $75-95.

Bristol Yellow #6733. Footed jar with applied black threading and black stopper. Steuben Glass. 6 1/2 inches. $500-550.

Three toed rose pattern covered dish with lid. 3 1/2 inches. Amazing pattern similarities to the U.S. Glass Rose pattern but no documentation for this specific shape found. $58-78.

Flat, ribbed covered dish with lid, 4 inches diameter. Unattributed. $22-32.

Queen Anne pattern powder box with lid. New Martinsville Glass. 4 1/2 inches diameter. $35-48.

Three toed ribbed covered dish with lid. Akro Agate Glass. 4 inches diameter. (Hardy 1992) $22-32.

Rose or flower blossom shaped figural powder dish with lid. 4 inches diameter. $40.

Trudy pattern #24 tumble-up or guest sets. Morgantown Glass. Two part, yellow bottle with black tumbler-lid. Circa 1930s. (Snyder 1998) $120-160.

Bottle or cologne. Ground and polished bottom, hand cut and faceted small stopper. This finely made and elegantly finished flat little bottle is probably of European origin and meant to be carried by a lady. 4 1/2 inches tall. $30-45.

Lobed covered powder box with lid #724. Westite Glass. Gold decoration. 1930s. (Kovar 2004) $45-60.

Trudy pattern #24 tumble-up or guest sets. Morgantown Glass. Water bottle 6 inches with 3 inch tumbler as lid. Two part. (Snyder 1998) $80-110.

Paneled Grape line #1881 round box with lid PG-34, Westmoreland Glass. (Kovar 2004) $20-28.

Below:
Bathroom fixtures, to be set into tiled walls. Westite Glass. Left to right: soap dish, $35-45; towel rod end holder, $20-25; bathroom tumbler, $35-45; toothbrush holder, $45-55. Manufactured 1930s. Note: Toothbrush holder embossed "Balmer-Westite Co. Weston, West Virginia Patent Pending." Tumbler embossed "Westite Pat'd Weston, West Virginia." See company history chapter for original illustration.

Cosmetic jar. Embossed "Tussy" on base. 4 1/2 inches. $8-12

Bath salt bottle with stopper. Pinched sides, 5 1/4 inches tall. Unattributed. $55-65.

Barbers bottle. Silver overlay decoration "Water." Metal dispensing stopper. Maker unknown. Bottom embossed "Terminal Barber Shops." $80-95.

Bottom of bottle "T BARBER SHOPS" showing embossing.

Beaded Medallion pattern tray. "Made in Czechoslovakia" paper label often found on piece. 9 1/4 inches. $26-32.

Chapter Three
Bowls and Baskets

BOWLS

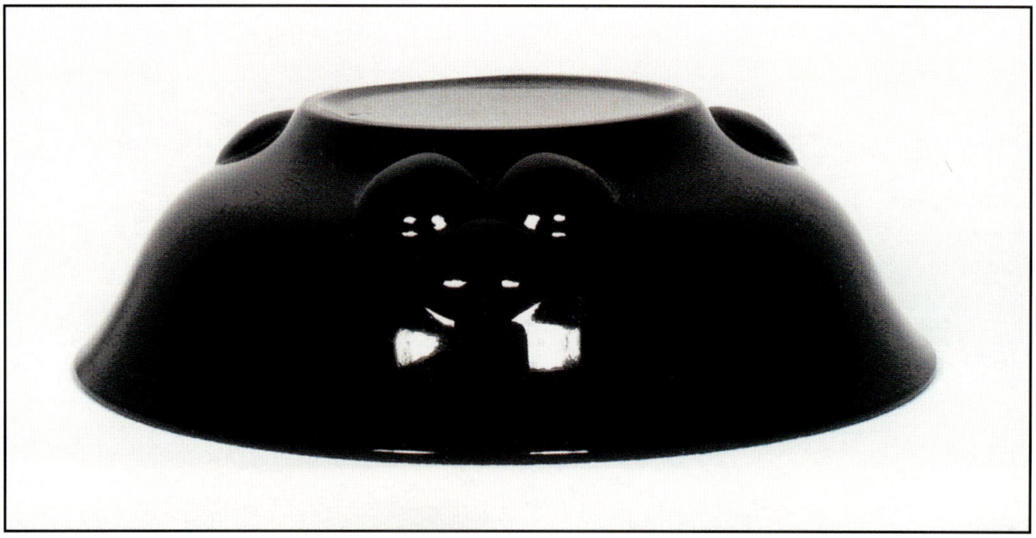

Three Ball bowl #1067-4. Westmoreland Glass. "Bell bowl" 11 3/4 inches ground bottom. Note that there are matching candlesticks and they are illustrated in the candlestick chapter. (Kovar 2004) $59-79.

Console bowl #9320 with base, gold bands decoration on satin glass. U.S. Glass. 9 inches diameter. Original sticker on base. Base is also U.S. Glass. Set as shown, $50-74.

Concord Bowl #1864. Pilgrim Glass. 9 inches. Cased glass, black over white or opal interior. The shape appears in Pilgrim catalog and the same time period catalogs featured colored case over opal (milk white) glass. (Form illustrated in Pilgrim 1997/B catalog.) $25-38.

Gold Brocade etched and gold encrusted rim bowl. Lotus Glass decoration on Unattributed low 11 inch bowl. $35-48.

Lilique Bowl #8142. U.S. Glass. Circa 1927. Satin. $45-75.

Shell formed bowl, gold trimmed. Unattributed. $18-28.

Whitehall Planter. Indiana Glass, from Fostoria's American pattern molds acquired by Indiana after Fostoria closed. American in any color is uncommon to rare. Whitehall is not. 6 x 4 inches. $6-10.

Mixing or batter bowl. Unattributed. $14-26.

Iridescent "stretch" glass bowl. $60-65; plate with ring for bowl. $65-75.

Low, shallow bowl with crimped edge. Unknown manufacturer. 12 inches. $20-28.

"Long Oval Bowl" #1562-2. Fenton Art Glass. 13 inches. (Whitmyer, 2003) $50-65

Rolled edge console bowl Party Line #191. Paden City Glass. 13 inches. (Domitz 2004) $50-65.

Console bowl Party Line #191 Paden City Glass. 12 inches, flat rim. Note that this and the rolled edge Paden City bowl nearby are from identical molds with the treatment given the rim while the glass was hot making the differences in the final bowl form. (Domitz 2004) $45-55.

Bowl, low straight sides, flat rim. Unknown manufacturer. 8 1/4 inches. $30-38.

Cupped bowl. Unknown manufacturer. Ground bottom. 8 inches. $20-26.

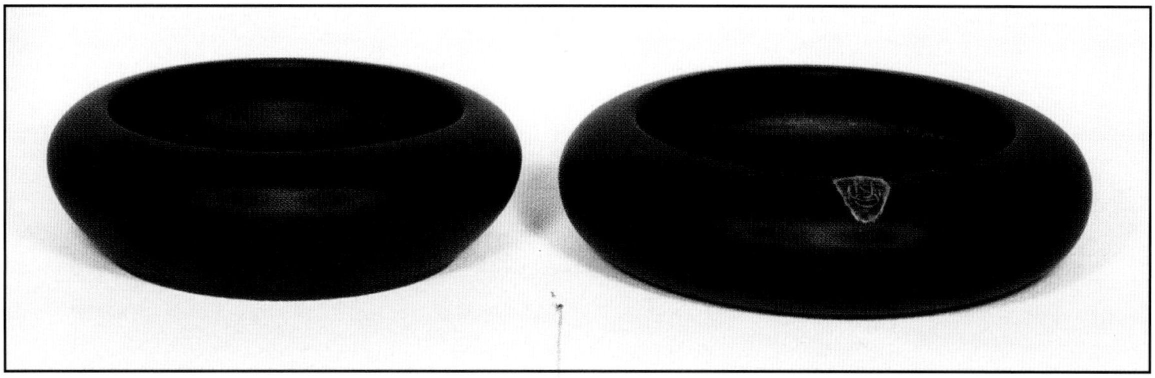

Floral bowls, U.S. Glass. Left: 6 inches diameter #15151; Right: 7 inches diameter #15151 with original manufacturers label. Note original paper label adds 10% to value. Either as shown, $22-34.

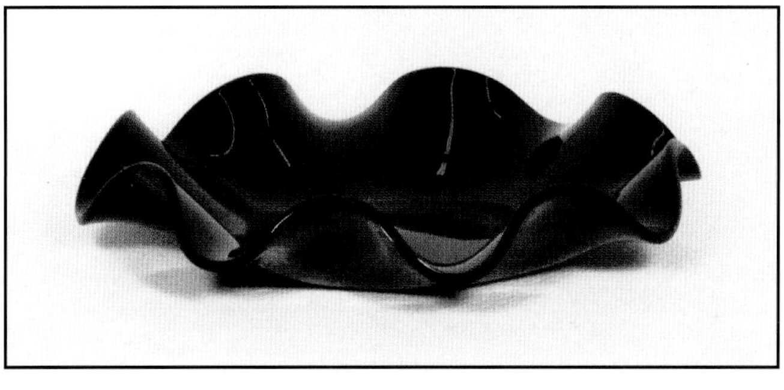

Low fluted edge bowl. Possibly Houze Glass. Slumped to attain form/shape. 7 1/2 inches. $12-18.

Oval bowl, shallow. Unknown manufacturer. Ground bottom with slight 1/4 inch recess in side circumventing entire bowl near base. $20-28.

Free Form bowl. 11 1/2 inches. #1800 Morgantown. $50-75.

Right:
Melon form bowl with small feet. Unknown manufacturer. Feet ground. 5 inches. $20-28.

Spin mold bowl. Dalzell-Viking. 6 inches. Note each of these will vary erratically due to the organic nature of the manufacturing process of spin molding. Dalzell-Viking era 1987-1998. $18-24.

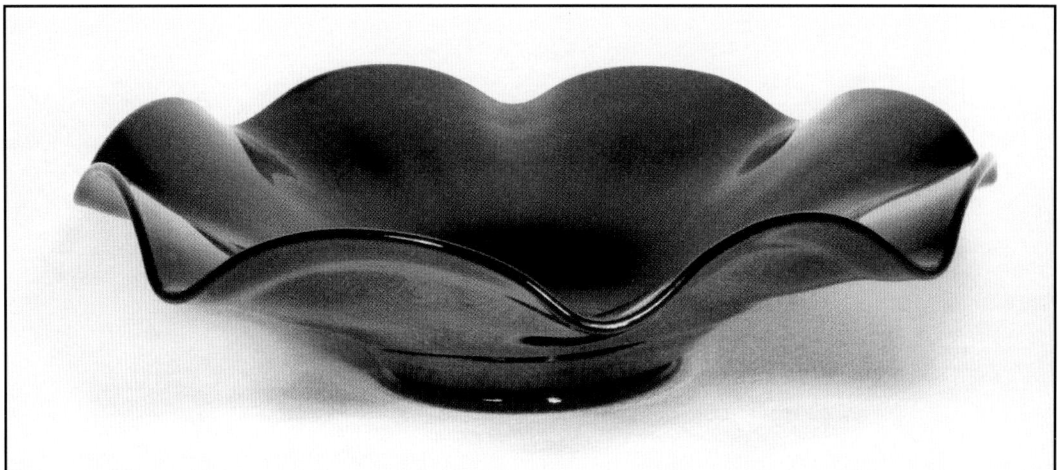

Console bowl, fluted. Unknown manufacturer. Bowls of this nature are so similar in form and shape with no distinguishable characteristics that attribution is always difficult and at times impossible. 12 inches. $20-28.

Console flared bowls. Diamond Glass. Left: enamel rim 10 inches; Right: undecorated. 9 1/2 inches. Either bowl, $20-28.

Console bowl #24. Cambridge Glass. Hand-painted floral, 10 inches. Circa 1925. Matching candlestick shown in candlestick chapter. $42-58.

Diamond Point pattern bowl. Indiana Glass. 9 1/2 inches. This specific piece has some problems in sitting level due to accidental slumping when manufactured. When marketed by Indiana's Tiara home party plan division this same line becomes the Cameo pattern in black. $10-18.

Bowl, rolled edge. Fenton Art Glass, hand decorated floral. 9 1/2 inches. Signed "Hand Painted by S. Lee." $40-46.

Lily Bowl in Diamond Block pattern. Imperial Glass. 5 1/2 inches. $28-36.

Lotus or flared bowl. Fenton Art Glass #848. 9 inches diameter. Note this mold was used to create a number of flared out, cupped in, and other forms, all beginning with the same shape before manipulated while hot glass into the various forms. Note there are matching candlesticks (Whitmyer 1996) $28-38.

Console bowl, low footed. Unknown manufacturer. 11 inches. $20-28.

Flared rim bowl. Beaumont Glass. 10 inches. Original paper label. Handblown glass, most of the black bowls we handled in preparing this book were pressed glass; this is blown. $24-32.

Cupped bowl. Unknown manufacturer. 23 ray base, smooth interior. 7 inches. $22-32.

Console bowl. Unknown manufacturer. Rayed bottom, paneled interior bowl design. 8 inches. $20-28.

Satin glass bowl and footed bowl/ compote. U.S. Glass. Both 9 inches diameter. Flat bowl #15179. $20-28; footed bowl #15179. $30-38.

Minton gold encrusted and etched decoration, rolled edge bowl. Unknown manufacturer. Possibly decorated by Lotus Glass. 10 inches. $45-68.

Flared rim bowl. Unknown manufacturer. Footed. 9 3/4 inches diameter. $22-32.

Flared bowl with silver plate attached metal base. Unknown manufacturer. 11 inches diameter. $20-28.

Oval, ribbed flared bowl. Unknown manufacturer. 11 1/2 inches. Note: This form and shape is nearly identical to Stangle Pottery Colonial pattern 31338. Imitation of forms between pottery and glass has occurred often over time and may at times indicate the use of the same designer, at others times it is simple imitation/copying. $28-38.

Desert Blossoms pattern round footed bowl. Indiana Glass. 8 inches diameter. $18-26.

Footed, flared bowl with gold decoration. Manufacturer unknown. Gold lattice decoration. 10 1/2 inches. $24-38.

Nut bowl with enamel decoration. #515 L.E. Smith Glass. Hand-painted. Footed, turned up edge. 9 inches. (Weatherman 1974) $30-38.

Footed scalloped rim bowl. L.E. Smith Glass. Same as bowl that follows except flared out. Note rays on base. $28-40.

Autumn Console bowl #23. McKee Glass. 8 1/2 inches, footed, circa 1934. (Weatherman 1974) $58-68.

Footed cupped scalloped rim bowl. L.E. Smith. Enamel floral decoration. Same as bowl above except cupped in. Note rays on base. $28-40.

Ribbed bowl, cupped. #847 Fenton Art Glass. 6 inches. 1930s. $30-40.

Ribbed bowl, footed and flared. Fenton Art Glass #857. 10 inches. (Whitmyer 1996) $30-40.

"Vase" bowl. #1000 L.E. Smith Glass. 11 inches. (Weatherman 1974) $30-42.

Three toed bowl #100 "Wide Rib" pattern. Fenton Art Glass. Rayed base, wide ringed bowl with crimp, 10 inches. 1907-1934. (Whitmyer 2003) $18-28.

Three toed cupped bowl #525. L.E. Smith Glass. Silver band decoration. 5 inches. (Weatherman 1974) $18-26.

Three toed, small base, narrow bands of rings with crimp, bowl. 7 inches. Imaged upside down to show details on bottom, top plain. Similar to Fenton's Wide Rib viewed from top, but very different from this angle. Unattributed. $14-22.

Below:
Three toed console bowl with lobed rim. Diamond Glass. Silver decoration. 12 inches. $80-90. Shown with candleholders #716 with same decoration. Each, $40-50. *WVMAG Permanent Collection*.

Three toed console bowl, 9 inches with floral decoration shown on left, $18-26. On right, #1022 console bowl, 6 inches, $14-20. Both L.E. Smith Glass (Weatherman 1974)

Mt. Pleasant pattern console bowl, L.E. Smith Glass. Three toed with flat rim. Note the distinguishing hourglass shape decorations of the Mt. Pleasant pattern around the feet that continue up to near the rim. $28-40.

Three toed console bowl with pressed pattern on rim. Unattributed. 11 1/2 inches. $28-38.

Three toed flared rim bowl with platinum/ silver decoration. $16-24.

Three toed bowl, open toes. Imperial Glass. Circa 1930s. Feet with gold trim. 7 1/4 inches. (Weatherman 1974) $24-36.

Footed low bowl #3400/48 Cambridge Glass. Crimped, 12 inches. $50-74.

Open work bowl. #310 U.S. Glass. Satin. 12 inches. (Tiffin Collectors 2004) $60-70.

Open edge bowl. #8076. U.S. Glass. Satin. 10 1/4 inches. (Tiffin Collectors 2004) $45-55

Open edge bowl, low and shallow cupped edge. 11 inches. Possibly Lancaster Glass Open Edge pattern? $38-50.

Open edge bowl, crimped edge. Plum Glass. Marked on base with "PG" in keystone. 9 inches. $14-22.

Open edge bowl, low and shallow flat edge. Possibly Lancaster Glass Open Edge pattern? 12 inches. $38-50.

Open edge low flat bowl with cupped rim. 8 1/2 inches and footed lace edge with short stem, 7 1/2 inches. Unattributed. $18-26.

Doric pattern open edge, oval bowl. Westmoreland Glass. 12 1/4 inches. (Kovar 2004) $30-36.

Salad bowl with etching gold encrusted, two open handles. Central Glass, probably decorated by Lotus or Wheeling Decorating. 10 inches, ground bottom. $55-68.

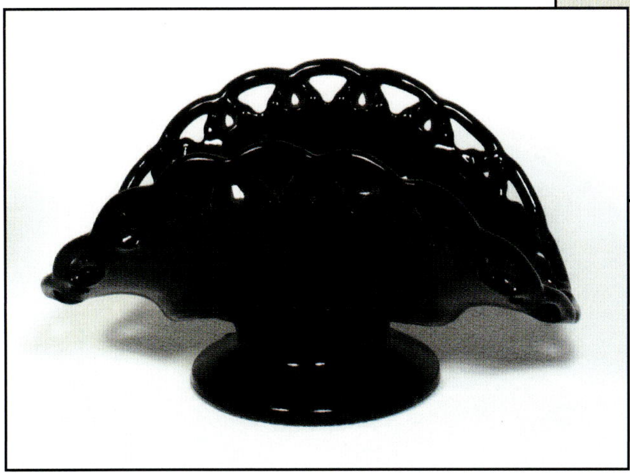

Open edge bowl, turned to create banana or fruit bowl. 9 inches in length. Unattributed. $48-60.

Low two handled bowl #2395. Fostoria Glass. (Long & Seate 2000) $85-100.

Mt. Pleasant pattern salad bowl #410. L.E. Smith Glass. Two open handles. 7 1/2 inches. (Weatherman 1974) $18-26.

Bowl, two open handles. L.E. Smith Glass. Silver decoration. 6 inches. $22-34.

Bowl, low with two open handles. Possibly L.E. Smith Glass due to similarity of form to known Smith pieces and handle shapes. 10 inches. $20-28.

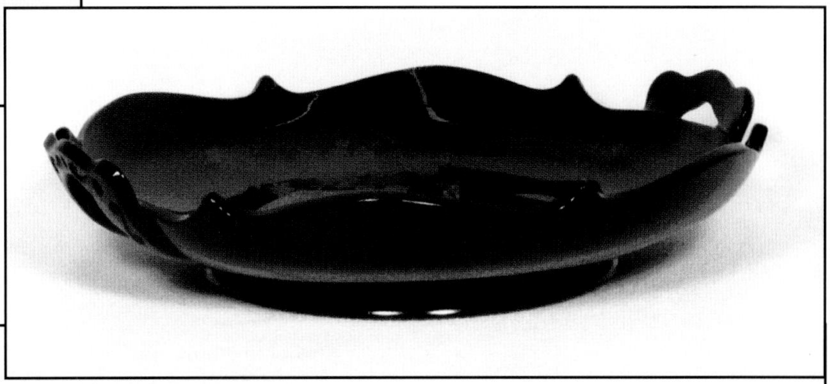

Dolphin handled bowl. #1621 Fenton Art Glass. 10 inches. Circa 1930s. $130-150.

Two handled fruit bowl #2400, 8 3/4 inches, $26-32 and handled footed bon bon #2400, $20-28. Both L.E. Smith Glass. (Weatherman 1974)

Console bowl. Libbey Glass. Black over crystal and then cut to clear. Shown sitting on a black square, the bowl has no foot or base. Circa 1920s. Acid stamp mark "Libbey" in circle. 4 inches tall, 10 1/4 inches diameter. $300-340.

BASKETS

Basket. L.E. Smith Glass. 6 inches wide. $24-34.

Paneled Grape Basket. L.E. Smith Glass. Unusual form. 5 1/4 inches wide. $30-42.

Basket. L.E. Smith Glass. 13 inches tall. $65-75.

Basket with crystal handle. Viking Glass. 5 1/2 inches wide. $22-34.

Basket, Janice Pattern with crystal handle. Viking Glass. Circa 1950s. 8 1/4 inches tall. $80-95.

Baskets, glossy finish and satin finish. #9574 U.S. Glass. Glossy 11 inches, Satin, 10 1/2 inches highlighting the variances in handmade glass, particularly hand applied handles! Either, $65-75.

Basket #6503. Fenton Art Glass. Footed, hand applied twisted handle with handle setter's mark. 6 inches tall. Made in 2003. $32-42.

Basket, Thumbprint pattern #4434 BK. Fenton Art Glass. 1968-1977. Hand applied handle with handle setter's mark. 8 inches diameter. $40-46.

Basket, small. 6 inches tall. Possibly Westmoreland Glass based on a similar, commonly found form. $18-24.

Basket, no visible pattern. Possibly Cambridge Glass. 5 inches tall. $55-68.

Chapter Four
Candlesticks, Lamps, and Lighting

Ring handled candlestick. 2 1/2 inches tall. Unknown manufacturer, but similar to #11032 made by Westmoreland Glass. In the Westmoreland version, the handle does not exceed above the top of the candle cup and there are other minor variances. (Felt and Stoer 2005) $20-24 each.

"Lifesaver" candlestick. #2446 Fostoria Glass. 1931-1935. 5 1/2 inches tall. $45-55 each.

Frog candlestick #72. U.S. Glass. 6 inches tall. 1924-1934. Satin. (Felt and Stoer 2005) $100-125.

Janice pattern candlestick. #4554 Viking Glass. 5 1/2 inches. Satin (Felt and Stoer 2005) $30-34.

Princess Plaza pattern candlestick. #5513, single light, Dalzell-Viking Glass. Late 1980s-1990s. Double light also made. 4 1/2 inches tall. Same stick with round base is Princess pattern. As shown, $24-30.

Two open handles candlesticks #408. L.E. Smith Glass. 3 1/4 inches tall. Decorated with white flecks. Circa 1950s. Note one stick of the pair is turned on side to show pattern under the base. (Felt and Stoer 2005) Each, $10-12.

Left:
Two open handles candlestick #1402. L.E. Smith Glass. 3 1/4 inches tall, 1920-30s. (Felt and Stoer 2005) $16-20.

Right:
Two open handled candlestick #408. L.E. Smith Glass. 3 1/4 inches tall. Circa 1950s. Note this is the same stick shown above but here without decoration. (Felt and Stoer 2005) $6-8.

Handled candlesticks. Attributed to Indiana Glass, circa 1920s. 4 1/4 inches tall. (Felt and Stoer 2003) Each, $10-12.

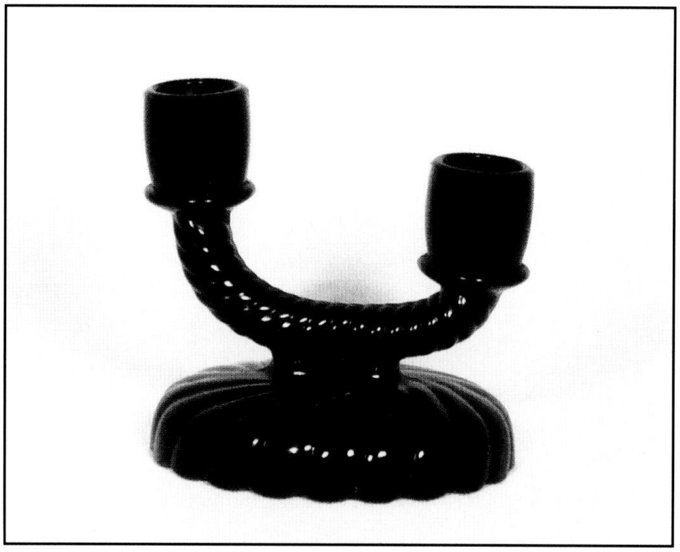

Two light candlestick. Imperial Glass. 4 1/2 inches tall. Introduced 1933. (Felt and Stoer 2003) $10-15.

Mount Pleasant pattern #600 two light candlestick. L.E. Smith Glass, circa 1930s. 4 1/2 inches tall. (Weatherman 1974) $24-30.

Cambridge Square pattern #3797/69. Cambridge Glass. Candle bowl, oblong. 7 1/2 inches. 1954-1958. (Felt and Stoer 2003) $40-50.

Trindle pattern three light #2383. Fostoria Glass. Produced 1927-1939. Can be found with etched decorations. (Felt and Stoer 2003) $35-40 each.

Three light U.S. Glass satin glass. Note a similar two light was also made by U.S. Glass but slight variations make it part of a named pattern, mid-1930s. $60-85 each.

"Keyhole" candelabra, three light. #638 Cambridge Glass. Made with decagon and round base. Also in company literature as #3400/638 6 inches tall. (Felt and Stoer 2003) $80-100 each.

Hobnail #3872 candle bowl. Fenton Art Glass. 6 1/2 diameter. Introduced 1968. $18-22.

Petal form three toed candlestick. #848 Fenton Art Glass. 1932-1936. 4 1/2 inches diameter. $10-16 each.

Mt. Pleasant pattern single candlestick. L.E. Smith Glass. 2 1/4 inches tall, circa 1930s. (Felt and Stoer 2005) $12-16 each.

Single light candlestick. Hazel Atlas Glass. Silver floral decoration. 2 1/2 inches tall, circa early 1930s. (Felt and Stoer 2003) $10-14 each.

Single light candlestick, shown with candle. Unattributed. $18-22.

Candlestick #93. U.S. Glass. 3 1/2 inches tall. Satin. 1929-1939. $15-20 each.

Candlestick. #99 Diamond Glass-Ware. Circa 1930-31. 3 inches tall. (Felt and Stoer 2003) $20-25.

Candlestick. L.E. Smith Glass. Silver decoration. 1930s. Add 10-20% if decorated. 3 inches tall. (Felt and Stoer 2005) $8-12 each.

Candlestick. H.C. Fry Glass. 3 inches tall. Introduced 1928 and probably made until 1933. Made with plain (as shown) and diamond optic patterned foot. (Felt and Stoer 2003) $20-25 each.

Candlestick #27. L.E. Smith Glass. Circa 1930s. 2 1/2 inches tall. Becomes #27/9 when silver decorated as shown. See F.W. Woolworth page in Smith history later in this book. As shown, $10-14 each.

Candlestick #2000. Central Glass. Etched decoration #62. Circa 1930s. Gold decorated. 3 inches tall. Each, $48-51.

Candlestick, large center disk. Diamond Glass-ware and/or Indiana Glass. Circa 1920s. 4 inches tall. (Felt and Stoer 2003) $10-26 each.

Candlestick Pillar Flute pattern #39. Imperial Glass. Paneled underside. 2 3/4 inches tall. 1925-1943. (Felt and Stoer 2004) $14-18.

Candlestick. Maryland Glass, showing shard of similar stick recovered from old factory site. Note this form is nearly indistinguishable from a similar shape produced by Fostoria Glass. $16-24.

Candlestick #2324. Fostoria Glass. 2 inches tall shown, manufactured in 5 sizes over time, as shown 1926-1932. (Felt and Stoer 2004) $15-20 each.

Candlestick, 8 panels. 4 inches tall. Unattributed. $16-24.

Spire pattern candlestick. #211 Paden City Glass. 5 inches tall, late 1930s. May be found with etched decorations. As shown, $30-40.

#2935 1/2. Fostoria Glass. Etch Fern decoration and gold. Circa 1929. (Long and Seate 2003) $55-70.

Leaf Candleholders. Indiana Glass for Tiara Exclusives in 1989. Candleholder created from Indiana's earlier #1009 Leaf Pattern sherbet. The original Indiana pattern was not produced in black. 3 3/4 inches tall. $12-15 each.

Winged candlestick. #16 Duncan & Miller. Circa 1930-1939. Commonly marketed with matching winged bowl. (Felt and Stoer 2003) $35-50 each.

#2395 1/2, 5 inches tall and right #2395 3 1/2 inches tall—both Fostoria Glass. Taller/left circa 1929-1939, $30-35 each; and right 1929-1930, $25-30 each. (Felt and Stoer 2003)

Wigwam pattern. L.E. Smith Glass and/or Co-operative Flint Glass. Attribution unsettled by other glass scholars at this time. 3 inches. (Weatherman 1974) $24-34.

Candlesticks #2402. Fostoria Glass. 4 1/2 inches diameter. Designed by George Sakier. Made circa 1929-1939. Each, $22-30.

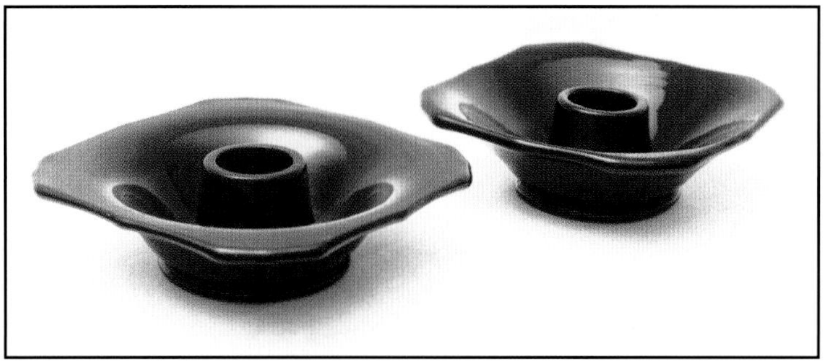

Mrs. "B" pattern #411. Paden City Glass. 1 1/4 inches high x 4 inches diameter. Made circa 1929 through 1930s. Each, $14-18.

Octagonal candlesticks #448. Viking Glass. 2 1/2 inches tall. Part of the Christina line of shapes. Produced 1945 to early 1950s. (Six 2003) Each, $10-14.

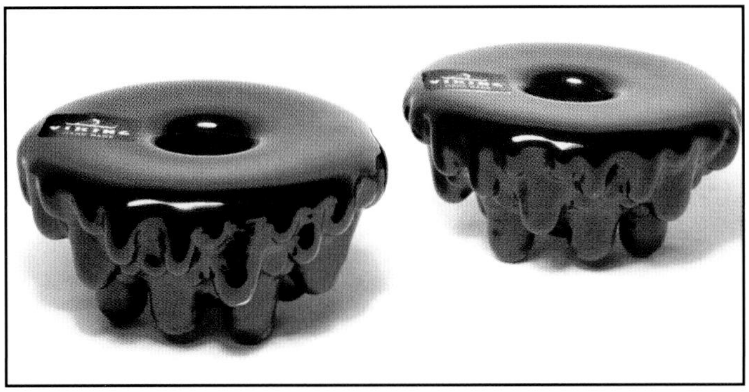

Teardrop pattern #7841. Viking Glass. 1 5/8 inches tall x 3 1/4 inches diameter. Made 1978-1980. Each, $8-12.

Worthington candleholder #4084. Libbey Glass. Circa 1994 to date. Mold originally Viking Glass #8423. Panel sided. 4 1/2 inches tall. (Felt and Stoer 2002) $6-8.

Crows Foot pattern #412. Paden City Glass. 2 3/4 inches tall. Each, $32-44.

Cornucopia candlestick #65. Dalzell-Viking Glass. 4 inches tall. $12-18.

Rolled edge candlestick. Unattributed. Circa 1920-30s. 3 inches tall. $18-24.

Petal based candlestick. #305 L.E. Smith Glass 2 3/4 inches tall. Circa 1950s. Not to be confused with the very similar Fostoria stick that sits flat on the base. Authors Felt & Stoer note this was later redesigned to add a sculpted petal-like top to the candle cup, thus becoming #4821 Simplicity. $8-12 each.

Mushroom top candlestick #718L. Imperial Glass. 1925-1931. 2 inches tall. $20-25.

Left candlestick Westite Glass, 1 3/4 inches tall x 3 1/2 across base. $22-30. Right candlestick U.S. Glass 1 3/4 inches tall x 3 across. $12-16. Note Westite stick nearly identical to ones made by Akro and Houze. Houze will have "851" on base. Akro will have single digit number while Westite has no number.

Octagonal base candlestick #94. U.S. Glass. 1926-1939. Enamel decorations. 1 1/2 inches tall. (Felt and Stoer 2004) As shown, $14-20.

Iridescent candlestick #2324. Fostoria Glass. Uncommon decoration "Onyx Luster." (Long & Seate 2003) $150-200.

Gadroon line candlestick. Fenton Art Glass. Originally Cambridge Glass #3500/74 mold but here marked Fenton under base. Circa 2000. 4 inches tall. $20-26.

Candlestick #1002. Smith Glass. 1920-1950s. 3 1/8 inches tall. (Felt and Stoer 2004) $10-14.

Centennial pattern #727. Cambridge Glass. Same form later referenced by Cambridge as #3400/627, Martha Washington and Victoria. 4 inches tall. (Felt & Stoer 2003) $24-34.

Tall candlesticks #1271. Cambridge Glass. 6 1/4 inches tall. Made circa 1922 to 1930. Same sticks also shown in following photo with bowl and gold decoration. Each, $65-85.

Tall candlesticks #1271. Cambridge Glass. Gold decoration. Circa 1920s. Candlestick. Each, $80-95; bowl, $55-65.

Tall candlestick #2000. Central Glass. Gold floral decoration. See matching console bowl in bowl chapter. As shown, $35-48.

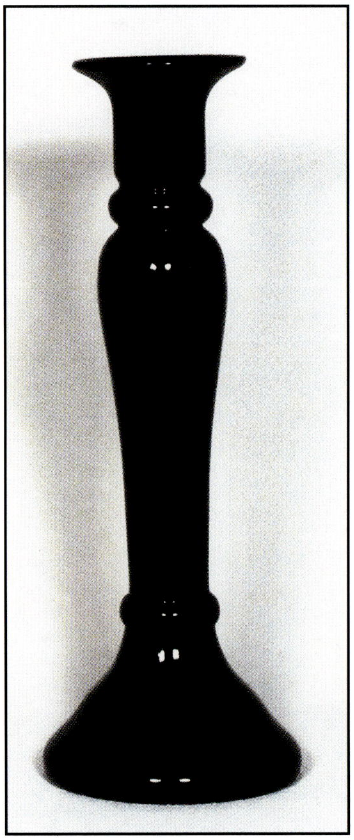

Blown candlestick. Diamond Glass. Circa 1925. Tall hollow/ blown. 9 1/2 inches tall. (Felt & Stoer 2003) $25-35.

Tall candlestick #70. U.S. Glass. 1922-1935. 6 1/2 inches with gold band decorations. Satin. $30-34.

Octagonal candlestick #4127. Dalzell-Viking Glass. 1993-1998. 5 1/2 inches tall. $10-16.

Square candlestick #200-5. Cambridge Glass. Shown with candle. Circa 1920s. $35-45.

Tall candlestick with floral enamel decoration. #116 Paden City Glass. 6 3/4 inches tall. Circa 1924-1930s. Felt & Stoer note, "This was one of several candleholders appearing in a 1929 advertisement by Thayer & Chandler, Chicago, as being suitable for hand painting by hobbyists." This may explain the decoration shown here? Similar to tall sticks made by Central, Lancaster, Westmoreland and others. See Felt & Stoer for more. (Felt & Stoer 2005) $40-50 each.

Square candlestick #3400/627. Cambridge Glass. One stick on side to illustrate the rayed bottom. 4 1/4 inches tall. Circa 1930s. Each, $34-38.

Twist candlestick #66. U.S. Glass. 8 1/4 inches. Satin. Each, $30-38.

Square tall candlesticks. 6 3/4 inches. Satin. With uncommon matching glass hot weather candles made by U.S. Glass/Tiffin. The candlesticks may be U.S. Glass/Tiffin #63, circa 1926, but cannot be attributed with certainty, since many other companies made similar ones. (Felt & Stoer 2005) Candlestick alone, $20-26. Candle, $95-100 each.

Tall candlestick #56B. U.S. Glass. Gold band decoration. Circa 1920s. $40-50.

Spiral candlestick #1139. Indiana Glass. Made 1990-1991. 6 inches tall. (Felt & Stoer 2003) Each, $10-14

Two candlesticks, iridescent/stretch glass. Left Northwood Glass #657, circa 1920s, 8 1/4 inches tall. $65-85. Right Imperial Glass #6009, 8 7/8 inches. $80-100.

Co-operative Flint Glass candlesticks. Left: $20-28; right: #332 Ray or Douglas shape, circa 1920s, $35-48. Attributions from glass researchers JoAnne and Earl Autenreith who are preparing a book on this company as we go to print with this book.

Maple Leaf pattern. Plum Glass #1928. Made in the 1990s from original Westmoreland Glass mold used in the 1930s. Plum version bears the Westmoreland logo in a circle with three slashes in the center. 5 3/8 inches tall. Each, $12-20.

Tall candlesticks #114. Paden City Glass. Panel sided. Made 1920s-1930s. 6 1/4 inches tall. Each, $20-28.

Barton pattern #9935 candleholder. Morgantown Glass. 4 inches, circa 1960s. $16-20.

Votive candle cups. Dalzell-Viking Glass. Both 3 inches tall. Made in spin mold so extreme variances in height and top finishes will occur. Either form. $6-12.

By Cracky pattern. L.E. Smith Glass #3133. 2 1/2 inches tall. Circa 1920-30s. One candlestick on side to shown pattern on bottom. (Weatherman 1974) $20-26 each.

Modernistic pattern #33. New Martinsville Glass. Introduced in 1928 and continued in production until the early 1930s. Design by Robert E. McEldowney, general manager of the company. 2 1/2 inches tall. An extensive line of other objects bear the same triangular shape. (Felt & Stoer 2005) Each, $30-35.

Monarch pattern. Indiana Glass #10426. Made 1977-1978 for Tiara Exclusives. This was a new mold created to supplement the reissue of one of the earlier Indiana patterns which originally did not have candlesticks, thus the look of an earlier style. 1 3/4 inches tall. One candlestick turned on side to show candle socket and interior. Each, $8-10.

Three Ball pattern candlesticks. Westmoreland Glass #1067. Each, $30-40. Shown with turned edge bowl #1802, $20-30. Candlesticks are 3 3/8 inches tall. Made in the 1950s.

Votive candle cup. Brooke Glass. 2 inches tall. Circa 1970s. $6-10.

Cambridge Square pattern #3797/493. Cambridge Glass. "Ebony," 2 1/2 inches tall shown here, three sizes were made. See original catalog illustration in company history chapter. (Felt & Stoer 2003) $22-34.

Cambridge Square pattern #3797/495. Cambridge Glass. "Ebony," 3 1/8 inches tall. (Felt & Stoer 2003) $22-36.

Tall frogs #69. U.S. Glass. Satin. See related pattern earlier in this section with shorter frog sticks. Each, $150-175.

Cathay Shen candle bases #5020/2. Imperial Glass. Made in 1982 and marked "ALIG." This is a reissue of the Cathay line made in crystal only from 1949-1953. 2 1/4 inches tall x 3 1/4 inches square. $45-55.

Swan candleholder #922. Viking Glass. 5 inches tall including crystal neck. Made circa 1950-60. $30-38.

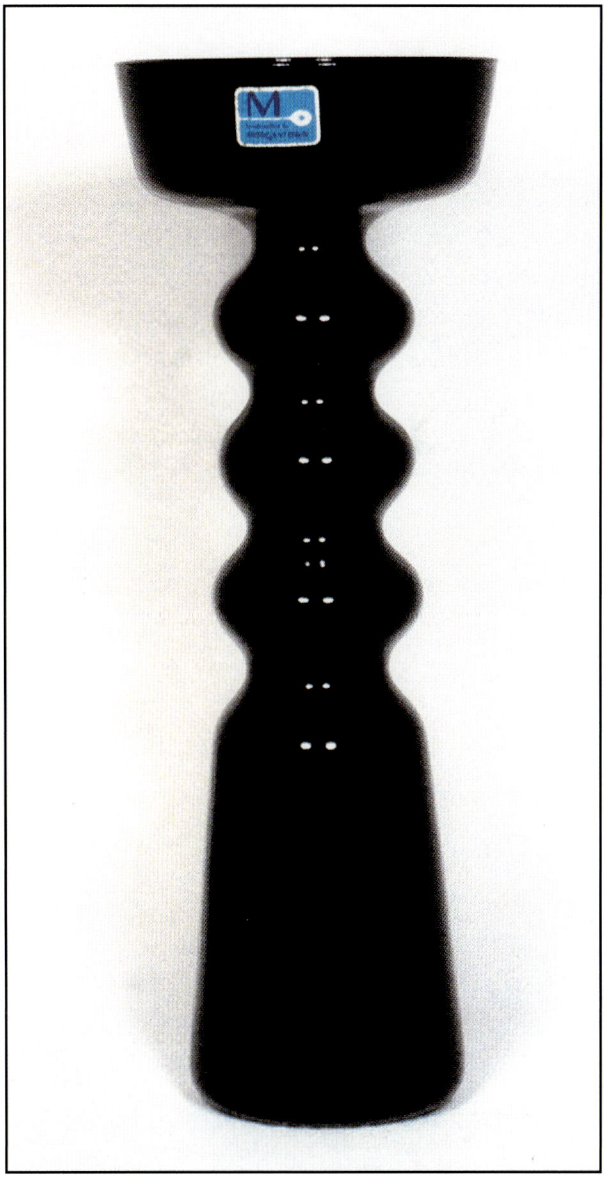

Spindle candlestick #1205. Morgantown Glass. 10 1/4 inches tall. (Snyder 1998) $38-44.

Electric lamp. Houze Glass. Black foot, marbleized/slag center, and black glass top, all drilled and wired. 9 inches. $65-85.

Electric lamp. Black base, crystal disk, and black glass "font," all drilled and wired. Circa 1930s. $28-45.

Coronet pattern candleholder #105. Morgantown Glass. 9 inches tall. 1963-1971. $20-32.

Left:
Scalloped base Onyx art glass lamp. Dalzell, Gilmore and Leighton Glass Co. Font and base threaded. Only base of this form known. Circa 1889. No value established.

Below:
Electric lamp base. Owl lamp. Cambridge Glass. As shown, $50-75.

Right:
Two Post Onyx art glass lamp. Dalzell, Gilmore and Leighton Glass Co. Font and base threaded. Lamp base in black found with other colored fonts. Circa 1889. No value established.

Chapter Five
Candy Jars, Cookie Jars, and Containers with Lids

Covered candy #7380. Fenton Art Glass. Marked Fenton in oval. 9 1/2 inches. This is a form probably made for the decorated lines Victorian Bouquet (1995) and Autumn Leaves (1993-94). Not sold as shown outside of the Fenton gift shop. $20-32.

Covered candy #5960 BK. Fenton Art Glass. 5 3/4 inches tall. Produced 2003. $40-46.

Covered candy with gold decoration. Maryland Glass Co. $24-38.

Covered candy. While the lid and base patterns seem different, we saw this combination repeatedly while preparing the book, convincing us that it was marketed as a jar and lid. Some sources have suggested this is a covered powder but we think the cone shaped bowl would make a candy dish use more likely. The patterns in the glass are known Hazel-Atlas patterns so we suggest that as an attribution. 3 3/4 inches wide. Circa 1930s. $12-20.

Covered candy #15179. U.S. Glass. Wide rim "bon bon." 10 inches. Gold decorations. Satin. $75-90.

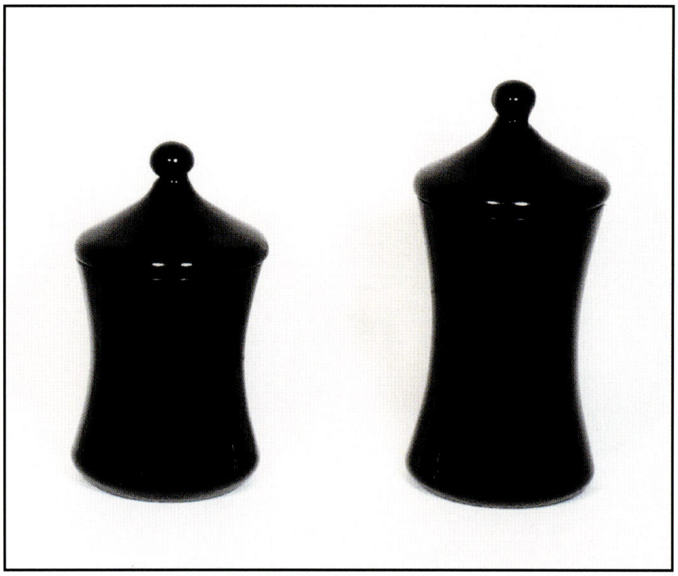

Two covered candies. Morgantown Glass. Left: 6 inches; Right: 7 inches. Right etched with "Reclamation Association West Virginia Surface Mining 1966-1976." Either size, $18-26.

Temple Jar #7588. Fenton Art Glass. Silver Poppy decoration on ebony. Produced 1981-1982. Signed "Hand painted by D. Jewel." 10 1/2 inches. (Walk 2001) $75-85.

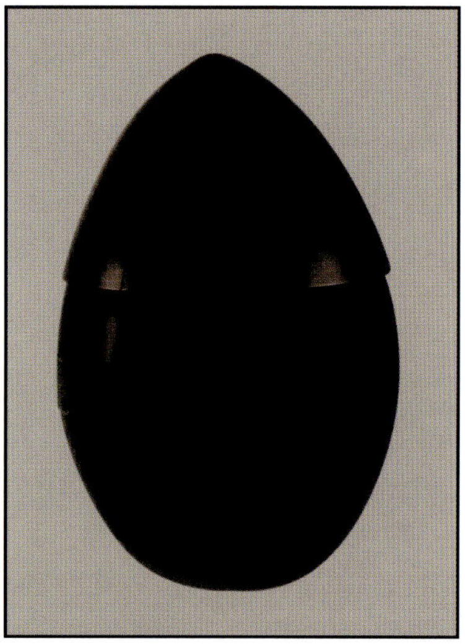

Eggcentric covered candy #1407. Morgantown Glass. Ebony base, metal lid. 5 1/2 inches. (Snyder 1998) $30-45.

Ginger Jar. Gold decoration. Satin. Unattributed shape and decoration. $55-65.

Two covered candies. Maryland Glass. Left: etched crystal bowl with black stem/base and lid, the etch is Springtime by Lotus Decorating. $90-100; Right: etched band near lid with silver decoration on etch and foot, black stem/base and lid. $45-60. Note the distinctive short stem has decreasing sided wafers, the largest having serrated edges like the sides of a coin or cog.

Ginger Jar. Gold decoration of butterflies. Satin. Unattributed shape and decoration. $45-55.

Covered candy. Unattributed manufacturer. 48 rayed base, gold band decorations. 6 1/2 inches diameter. $30-38.

Wave Crest pattern candy box #6080. Fenton Art Glass. Copper Rose floral decoration. "Hand painted by V. Gherke" on bottom. 4 3/4 inches tall. Circa 1990-1991. (Fenton company catalog.) $75-85.

Candy/cookie. "Beanpot" with lid. Silver decorations. Possibly L.E. Smith Glass but unable to make positive identification. 7 inches diameter. $35-48.

Paneled Candy with lid #7108 BK. Fenton Art Glass. 6 1/4 inches tall. Circa 2003. $40-45.

Oval candy #319. U.S. Glass. Enamel decoration. 8 1/2 inches with lid. Circa 1920s. $65-80.

Cookie jar with lid. L.E. Smith Glass. 8 inches tall. Enamel floral decoration. Decoration does not alter value. See company history chapter for illustration. $60-75.

Cookie jar #3. L.E. Smith Glass. 8 1/2 inches tall. Note recessed lid. See company history chapter for illustration. $75-90.

Macroon pattern #1681 covered candy. Fenton Art Glass. Collectors call this pattern "Big Cookies" at times. Wicker handle. 7 inches. Circa 1933. $90-120.

Cookie jar #1. L.E. Smith Glass. Barrel shaped jar with pointed handle. Possibly alternate or remodeled lid? $60-75.

Diadem pattern covered candy #2430. Fostoria Glass. Note small band of diamonds at base. (Long and Seate 1999) $45-60.

Casual line covered candy. Canton Glass, circa 1951. 6 1/2 inches diameter. Amber and black combination. (*All About Glass* Vol. 1 No. 1.) $58-66.

Round candy box with Fleur-de-lis finial. #2331 Fostoria Glass. Circa 1926. Decoration similar to several documented ones, but a variation. Note see similar Central glass form. As shown, $85-95.

Three lobed candy box with lid #106. Duncan & Miller Glass. Unidentified sterling silver decoration. Circa 1931. (Weatherman 1974) $45-65.

Round candy box with Fleur-de-lis lid. #2005 Central Glass. Decoration elaborate etching and gold encrustation. Maybe a decoration by decorating firm on a Central blank? Circa 1930. (Schmidt 2004) $85-95.

Chapter Six
Compotes and Stemmed Bowls

Dolphin handled footed compote #1608. Fenton Art Glass. 10 1/2 inches. Circa 1930s. (Whitmyer 2003) $85-120.

Mrs. B pattern oval footed bowl/compote #411. Paden City Glass. 10 1/4 inches. (Domitz 2004) $75-85.

Lotus footed compote #192. Westmoreland Glass. 4 1/4 inches tall. (Kovar 2004) $28-34.

Footed compote. Unattributed manufacturer. Eight point rim. 6 1/2 inches. $30-45.

Footed Touraine Bowl/Compote #15179. U.S. Glass. "Ribbon Black Satin" decoration. 10 1/2 inches. (Tiffin Club 2004) $55-75

Twist Stem compotes. U.S. Glass. Left: #315, 5 inches tall. $38-44; right: #3315, 7 inches tall. $42-50.

Thumbprint pattern #4425. Fenton Glass. Produced 1970-1974. Double crimped top. 7 1/4 inches. (Walk 2001) $28-34.

Comport. Fostoria Glass #2327. Onyx Luster – Mother of pearl iridescent on ebony glass. 7 inches. Produced circa 1926. (Long & Seate 1999) $120-149.

Compote #150. Cambridge Glass 10 3/4 inches. Rays under bowl and distinctive stem. Circa 1920s. $48-54.

Compote, fluted rim. Probable manufacturer is Indiana Glass. White decoration popularized by multiple companies circa 1950s. 6 3/4 inches tall. $28-36.

Fern etch decoration comport #2400. Fostoria Glass. The Fern decoration is plate etching #305, with gold added it becomes decoration #501. 6 inches. Produced 1929-1932. (Long & Seate 1999) $65-75.

Interior detail of bowl of Fostoria Fern etch comport.

Decagon pattern compote #1090. Cambridge Glass, 7 inches. (Cambridge Collectors 1976) $54-64.

Cumberland pattern covered compote. Morgantown Glass. Black glass stem, bowl, and lid with green finial and foot. 7 inches tall. Part of the "two tone Nubian line" circa 1920s. (Snyder 1998) $500-550.

Openwork comport #310. U.S. Glass. 11 3/4 inches. $58-68.

Draped Lady stem covered compote. Possibly L.E. Smith Glass. 9 1/2 inches tall. This reportedly was sold as a powder jar. However, when the contents are gone what remains is objectively a covered tall compote. We list it here due to the likelihood of seekers most likely to call it a compote. $90-110.

Chapter Seven
Flower Bowls, Florist Items, and Flower Frogs

Wallpocket. Pressed crackle pattern nail/ hanging hole on flat reverse. Possibly Smith Glass due to similarity to By Cracky line. 6 inches. $24-38.

Early American Hobnail pattern ivy ball #742. Imperial Glass. 6 1/4 inches tall. (Weatherman 1974) $28-38.

Handled little pot. Unattributed. 3 footed, two open handles. 5 1/4 inches. $14-22.

Three toed flower pot. McKee Glass. "V" patterned. 5 1/2 inches. See original illustrations in the company history chapter. $22-32

Golf Ball pattern Ivy Ball. Morgantown Glass. Crystal stem and foot, black bowl. $120-145.

Three toed flower bowl and frog. Beaumont Glass. Silver decoration. $65-75.

Three toed flower bowl and crystal flower frog. Unattributed. $30-42.

Floral bowl with frog #1012. L.E. Smith Glass. 6 1/2 inches diameter. (Weatherman 1974) $30-42.

Hobnail flower bowl and frog. 6 1/2 inches diameter. Possibly Smith Glass. $30-42.

Greek Key Band floral bowls. All L.E. Smith Glass. Left: 4 1/4 inches height with frog, silver decoration, $22-32; center/largest: 5 1/4 inches height with silver decoration, without frog, as shown, $14-18; right: 3 1/2 inches height without frog, embossed bird pattern on side, $24-38. Note please see original Smith advertisement in the company history chapter showing the Greek Key 3-footed jardiniere without a flower frog. We are uncertain if they were marketed with frogs but do appear with them on today's market.

Examples of some of the numerous styles of black glass flower frogs found. Today, with the passage of time, many are found without bowls or in use with bowls that they may not have originally been marketed with. *Note this study collection, from the Doris and Roy White collection, is property of WVMAG and available for closer scrutiny.* $8-20 each.

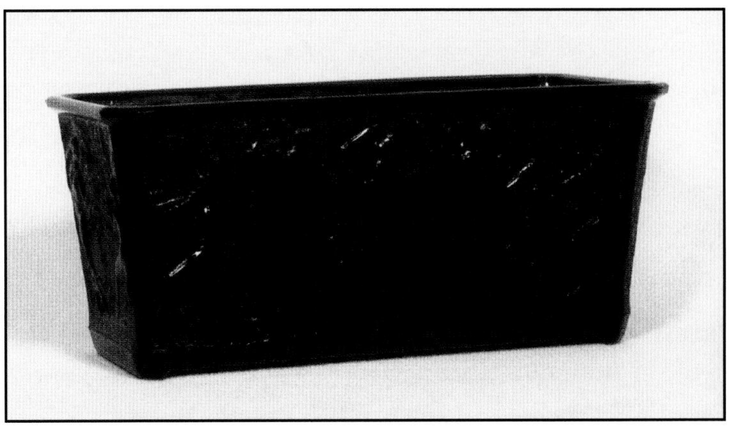

Dancing Ladies pattern Window Box #405. L.E. Smith Glass. 7 3/4 in length. Embossed on bottom in oval "L.E. Smith Co." and beneath oval "Made in USA." (Weatherman 1974) $38-48.

Window Box #2373. Fostoria Glass. Sterling silver overlay of impala. 8 1/4 inches in length. Decoration by unknown firm. $220-250.

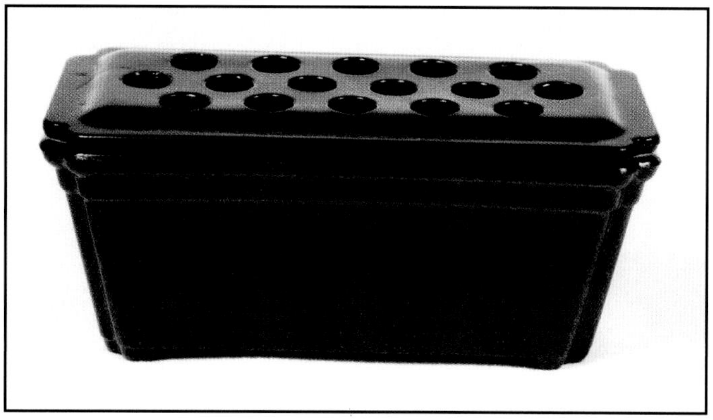

Window box with frog. L.E. Smith Glass. $30-45.

Window Box #2373. Fostoria Glass. With original frog. (Long & Seate 2004) $120-160.

Flowerpots. Westite Glass. 2 inches tall on left and 2 1/2 inches tall on right. Westite pots produced are as follows: Line #299, 2 inches; Line #300, 2 1/2 inches; Line #301, 3 5/8 inches; Line #302, 5 inches; Line # 303, 6 inches; and Line #304, 7 inches. As shown here, either size. $30-50.

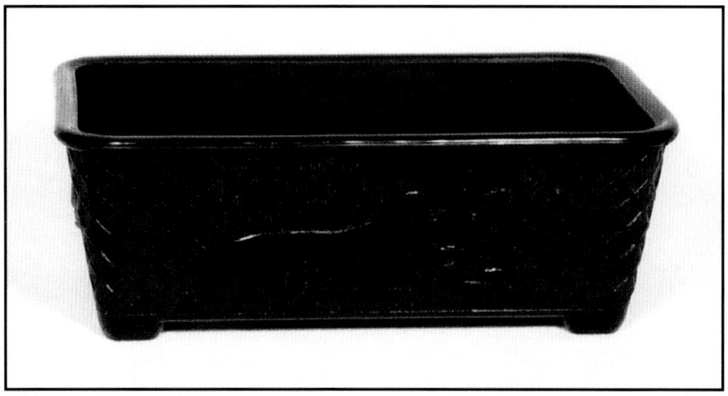

Lion and Lattice patterned Window Box. McKee Glass. See illustration in company history chapter. 9 1/4 inches length. $35-50.

Oval flower bowl. Bennett Glass. Base marked "B" in a triangle and "Made in USA 654" From altered original Akro Agate mold. 6 inches. Note: glass is a green/amber not amethyst under close inspection. $10-16.

Flower bowl. Westite Glass. Three toed base marked "Westite Weston W. Va." 5 1/4 inches diameter. $50-65.

Flower bowl. Westite Glass. Flared. 6 inches. $40-60.

Right:
Urn planter. 6 inches. Possibly Smith Glass, Unattributed. $10-16.

Chapter Eight
Kitchen Glass

Rooster egg cup. Unattributed. 3 1/2 inches tall. $22-32.

Flour Shaker. Manufacturer unknown. White enamel fleck decoration, plastic shaker top. Base marked "15 A 6" with letter A in a circle. 5 1/2 inches tall. $20-32.

Three ring mixing bowl. Manufacturer unknown. 7 1/4 inches. $20-28.

Acorn salt shaker "Congo #547." Buckeye Glass. Circa 1891. See trade quotes in the company history chapter attributing this form and color. We believe this is the first in print and documented attribution of this well known form and will lead to attribution of related colors in this same form by others.

Sugar Shaker. McKee Glass. Aluminum top, side decorated "Sugar." 5 inches tall. (Mauzy 2004) As shown, $20-34.

Salt and Pepper shakers. McKee Glass. Decorated "S" and "P." 4 1/4 inches tall. (Mauzy 2004) As shown, $30-45.

Block and Star pattern #5006. Fenton Art Glass. 1955-65. Metal plated chrome top. Each, $16-20.

Salt shaker, ringed base and foot, aluminum top. The rings are reminiscent of Fenton glass. $8-14.

Tall salt and pepper mill. Unattributed. Each, $16-22.

Salt and pepper shakers #210. Regina pattern. Paden City Glass. (Domitz 2004) $20-28.

Salt and pepper shakers. Silver decoration style leads to suggesting Smith Glass as a possible manufacturer. Each, $8-14.

Syrup and batter jugs with tray and covers. Paden City Glass. Circa 1936. (Domitz 2004) Set as shown, $100-140.

Egg cup. Unattributed. Black is single egg side, clear double egg side. 4 1/2 inches tall. $6-10.

Chapter Nine
Smoking Items

Far left:
Terrier #1 dog ashtray. L.E. Smith. Circa 1930s. (Weatherman 1974) $28-42.

Left:
Elephant ashtray. Unattributed, but has many characteristics of Smith Glass. Circa 1930s. $25-35.

Far left:
Scotty dog ash tray with match holder. Unattributed. Circa 1930s. $40-45.

Left:
Elephant #2 ashtray. 4 inches diameter. L.E. Smith Glass. Circa 1930s. (Weatherman 1974) $25-35.

Bull Dog ashtray. L.E. Smith Glass. 5 1/2 inches wide. 1930s. (Coe 2003) $35-45.

Black glass base ashtray with stylized metal deco duck. Unattributed. Circa 1930s. $40-60.

Black glass base ashtray with painted metal duck. Circa 1930s. Note several other metal attachments were used on black glass bases and most are in the same general price range. Some, such as skyscrapers or art deco themes, may go higher. As shown, $20-35.

Common round ashtray. A number of strikingly similar utilitarian ashtrays in black were viewed when gathering images for this book. Without more defining characteristics they are all in the $3-8 range and attribution seems futile.

Cockerel design #100 ashtray. L.E. Smith Glass. 5 inches diameter. Circa 1930s. (Weatherman 1974) $12-22.

A second common round ashtray, see similar_and caption. $3-8.

Ashtray #436. Viking Glass. Circa 1954 in ebony. Popularly called a fish shape ashtray. (Six 2003) $12-16.

White Daisies #4469 ashtray. Fenton Art Glass. 1972-1974. 6 1/4 inches diameter. (Walk 2001) $25-35.

Sphere ashtrays. Viking Glass. Left: 4 inches #6625, $10-14; and right: 6 inches #6827, $24-38. See the Viking catalog 1978-1979 included in the company history chapter.

Contemporary round, advertising ashtray. These can be found in a variety of similar forms with diverse advertisements. They would make an inexpensive and interesting collection. $2-6.

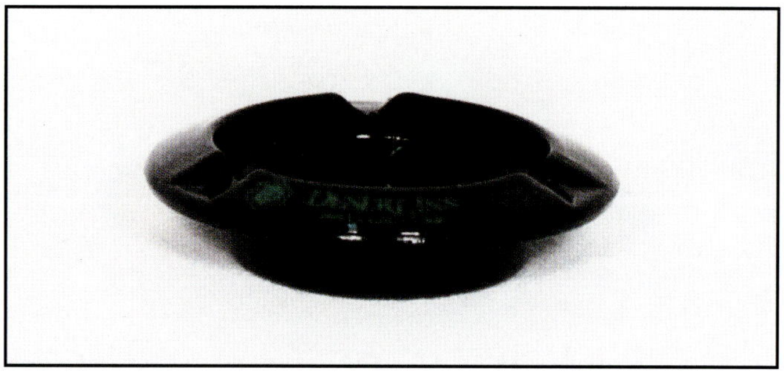

Contemporary round advertising ashtray, see similar_and caption. $2-6.

Stacked set of three ashtrays. Duncan and Miller Glass. Circa 1930s. Each, $4-12 depending on size.

Cambridge Square ashtray #3797. Cambridge Glass. Shown left inverted to illustrate the square base that is the repeating feature of the pattern. Right shown as it would be used. $18-28.

Three toed ashtray with sterling decorations. Fry Glass. (Fry Society 1990) $20-30.

Pilot ashtrays, shown stacked set of three. Viking Glass. 4 inches #775. $6-10; 6 inches #774, $8-14; 8 inches #773. $14-26. Illustrated in 1950s Viking Glass catalogs. (Six 2003)

Three toed "Club ashtray." Duncan & Miller Glass. (Weatherman 1974) $10-16.

Three Leg Round ashtray. #90 L.E. Smith Glass. 5 inches diameter, circa 1930s. (Weatherman 1974) $25-38.

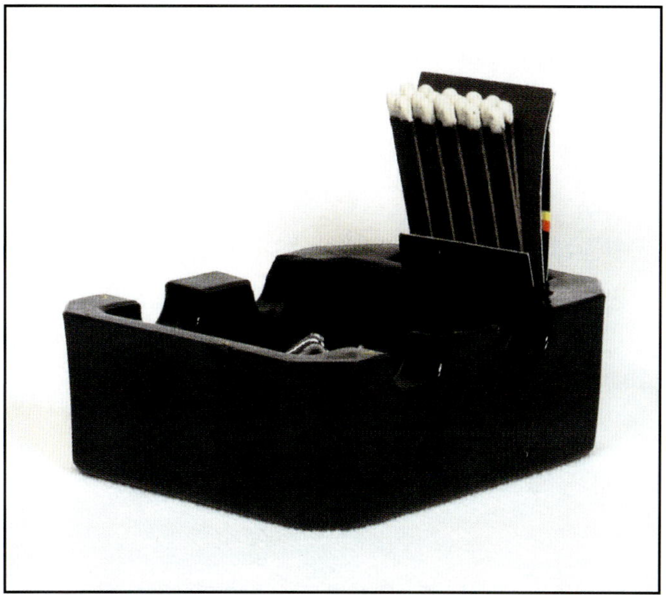

Rectangular with four cigarette rests and match book holding groove. Westite Glass. Circa 1930s. $18-28.

Clover. Hazel Atlas Glass. Match box holder in center. Circa 1930s. (Mauzy 2004) $60-90.

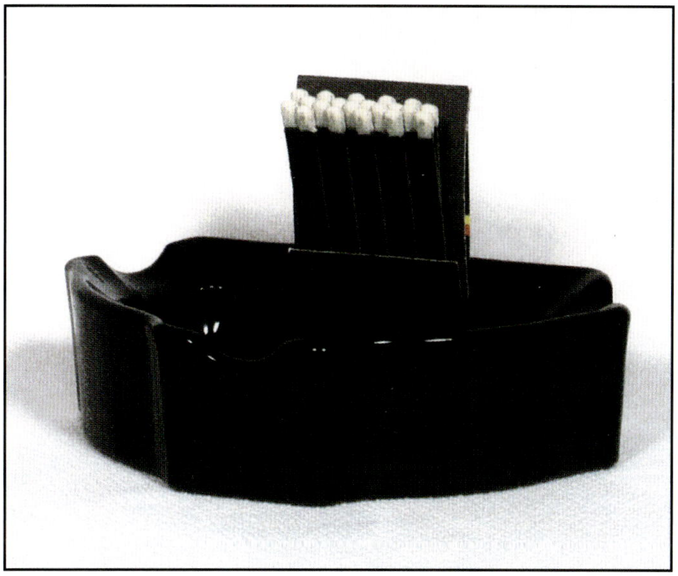

Triangular shaped with three cigarette rests and match book groove. Akro Agate Glass. 4 3/4 inches. (Hardy 1992) $18-26.

Advertising ashtray with match book holder in center. 6 inches. "American Legion Convention 19 Parkersburg 43." Unattributed. $14-16.

Rectangular with match box holder. 4 1/2 x 3 1/2 inches. Unattributed. $12-18.

Stag ashtray. #9298 U.S. Glass. Satin finish. (Tiffin Club 2004) $45-58.

"1902-1928 CARNEGIE NATIONAL BANK." #9298 ashtray. U.S. Glass. Oval 6 inches. Shown with satin finish, found both satin and glossy finishes. Satin about 10-20% more. As shown, $45-55.

"The Old Army Game" ashtray or pin dish (note no cigarette rests). Front shows male and female figures. Clothing styles early twentieth century. See other image.

Uncle Bucky's Ash Tray in the #2010 Frances line. Central Glass. Two part with removable lid that confines the smoke and has two conical depressions for snuffing the cigarette and two horizontal rests to hold cigarettes. A large hole in the center is for disposal of the smoked butt. Note the pattern inside the base is Central line #2010, Frances. Unusual decoration. $125-150.

Uncle Bucky's Ash Tray with lid off.

Reverse of "The Old Army Game" showing Gentleman's (?) hand lifting the lady's skirt. 5 1/2 inches. $60-80.

Uncle Bucky's Ash Tray showing interior Central Frances pattern.

Top Hat ash tray #9229. U.S. Glass factory D. Satin finish. 2 1/2 inches tall. Illustrated in catalog of 1926. $100-120.

Dobbs Top Hat ash tray. Embossed "DOBBS" on rim. Dobbs was a manufacturer of men's hats. 2 1/4 inches. $24-30.

Coal Scuttle ash tray #9386. U.S. Glass factory D. Illustrated in catalog of 1926. 4 3/4 inches long. Similar form produced by various manufacturers. Metal bail/ handle. (Tiffin Club 2003) $12-18.

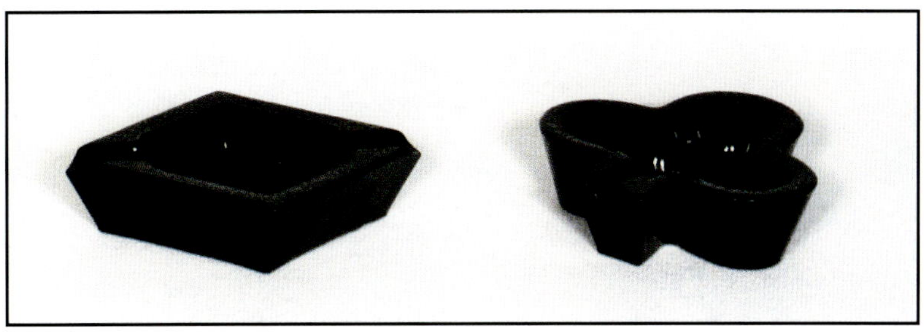

Card Suit ash trays. Imperial Glass in Heisey Glass molds. Diamond 4 inches in length; club 3 1/2 inches wide. All four suits are represented in a set of these. Each, $12-18.

Daisy and Button pattern three legged kettle ash tray. Widely produced and reproduced in various colors by Kemple Glass and others. Uncertain age. Unattributed. 3 3/4 inches diameter. $4-10.

Six sided ash tray. Bottom embossed "Weston W. Va. Made in USA" with the Westite logo. 3 inches, gold decoration worn. $18-28.

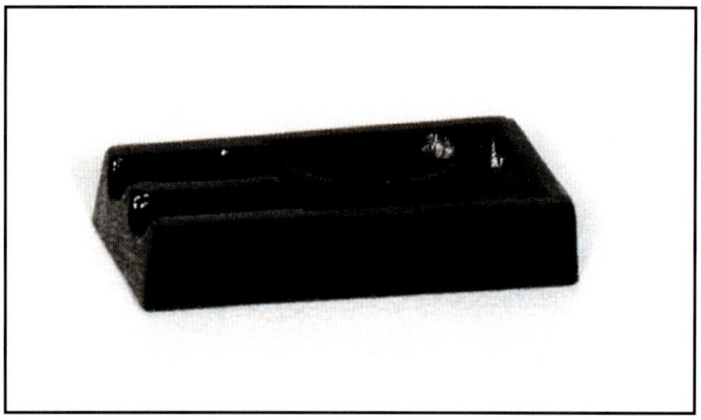

Rectangular ash tray. Westite Glass. Two long cigarette rests. Sold as a bridge set with black, pink, green, and milk ashtrays. 3 inches. 1930s. As shown, $18-22.

Square ash tray with four cigarette rests. #222 Akro Agate Glass. Marked on bottom "MADE IN US" with embossed logo of a crow flying through the letter "A." 3 inches. (Hardy 1992) $10-14.

Cigarette holder on stemmed base. Co-operative Flint Glass. 4 1/4 inches tall. $24-34.

Cigarette holder on stemmed ash tray base #2354. Fostoria Glass. 1926-1927 only. (Long & Seate 2000) $45-55.

Cigarette box with clear glass lid and two stacking ash trays. #1713. Cambridge Glass. Circa 1950s. See company history chapter. $100-130.

The previous cigarette box and ash trays revealing how they open.

Coaster and ash tray. 5 inches. Unattributed. $10-18.

Cigarette jar with lid. Westite Glass. 1930s. Gold decorations "Rotary International" with logo on lid. 3 1/2 inches tall. $45-75.

Cigarette jar with ash tray base. Only the base is glass, the balance is all metal. Base 5 1/2 inches diameter. Note the cigarette dispenser is illustrated partly lifted from the jar for the photograph. $20-30.

Cigarette jar with ash tray lid. Lid has three cigarette rests and doubles as ash tray. Vertical ribbing on cigarette jar. 4 inches tall. Often attributed to Hazel-Atlas but unconfirmed. $26-34.

Ash tray with metal fittings. 4 inches diameter. Possibly the metal was assembled and marketed by a firm who hired the glass produced for them (private mold work), and the glass manufacturer may never be discovered. The same applies to other pieces with other material added to the glass. $12-18.

Smoker set, cigarette urn, and lighter. Tiara Glass. Urn: 3 inches tall. Shown in 1978 Tiara catalog. Was sold as set along with 4 ash trays item #10001. Urn, $12-14; lighter, $18-26.

Ash tray, rectangular with three cigarette rests in front and match book holder on other side. 4 1/2 inches in length. Unattributed. $10-16.

Cape Cod pattern cigarette lighter #1602. Imperial Glass. Produced 1962-1966. 3 3/4 inches. (Garrison 2003) $40-50.

Thumbprint pattern cigarette lighter. #4495-BK, Fenton Art Glass. 2 1/4 inches tall. July 1968-1970. (Whitmyer 2001) $28-34.

Chapter Ten
Stemware and Barware

Carlton Etching #778 on Morgantown Glass #7606 1/2 Athena stemware line. Circa 1931. Goblet 7 1/2 inches. (Snyder 1998) $85-95.

Baden Etching #777 on Morgantown Glass #7653 Cantata line. Circa 1931. Twist stem saucer champagne (Baker & Kennedy 2003) $95-110.

Etching #425 on Central Glass stemware line #1448 goblet. Design Patented 1937. 7 inches. (Schmidt 2004) $30-38.

Faun Etching on Morgantown Glass #7640 Art Moderne line. Circa 1931. Sherbet 5 1/2 inches. (Snyder 1998) $45-65.

Fontinelle Etching #781 on Morgantown Glass #7620 Fontanne line, filament stem of black strip encased in crystal. Circa 1930s. Shown is liquor cocktail. 5 1/4 inches. (Snyder 1998) $110-125.

Artichoke line from Seneca Glass stem #1985. Black base, crystal stem and bowl. Goblet shown 6 1/4 inches. Produced in several other colors. (Page & Frederiksen 1995) $22-26.

Etching on crystal bowl, black stem and foot. Maryland Glass Co. Shown with black glass tray. No value established.

Cumberland Maryland glass product. 5 1/4 inches wine. Distinctive stem with unknown gray cutting. $12-18.

Other shapes of Maryland Glass etched stems shown on black tray. Etched. Circa 1930s.

Fostoria stem #4020. All crystal bowl with black stem and foot. Left to right: Millefleur cutting #195 (1929-1934) low sherbet 3 1/2 inches, $24-30; Fern Etching #305 goblet, $35-45; Millefleur cutting juice, $22-30; #4020 undecorated goblet (1929-1940) 6 inches, $30-38. (Long & Seate 2004)

Biscayne etching #747 on Morgantown Glass stem #7614. Shown is a wine, 5 1/2 inches. (Gallagher 2001) $48-56.

Byzantine etching on black stem and foot #15037. Tiffin/U.S. Glass Goblet. 8 1/2 inches. (Page & Frederiksen 1994) $52-62.

Intermezzo line of black encased crystal. Orrefors of Sweden, circa 1983. Tumbler, 6 inches, $38-48; goblet, 9 1/4 inches (www.replacements.com), $40-50.

Precedence - Onyx stemware line #6108. Fostoria Glass. Shown is claret 6 1/4 inches. (Long & Seate 2004) As shown, $45-50.

Topaz bowl and black foot and stem. Sloan Glass. The stems are three sided and the bowl has a vertical optic ribbing. Circa 1930-1932. Left: cocktail, $25-40; right: footed tumbler, $55-65.

Amber bowl, black stem and foot. Line #2007. Central Glass. Circa 1920-30s. (Schmidt 2004) Goblet, $25-38.

Frosted crystal glass bowl and black base. 4 1/4 inches, slant top beverage glass. Unattributed. $6-10

Morgantown crystal slanting bowl and black foot, 3 1/2 inches tall. $18-24.

Odd Ball #3010 milk glass base and black slanting bowl. Morgantown Glass. 4 inches tall. $55-65.

Clico pattern McKee Glass jade bowl and black square foot. 3 3/4 inches tall. See original company illustration in company history chapter. $26-34.

Eight Ball pattern. Morgantown Glass. Left and right glass 4 inches. Each, $45-55; center snack/pretzel bowl 8 inches tall, (Snyder 1998) $100-130.

Elizabeth pattern sherbet, Fenton Art Glass. 3 inches tall. 1930-1933. (Whitmyer 2003) $30-45.

Contrast stemware line #6120. Fostoria Glass. Sherbet, 5 5/8 inches. (Long & Seate 2003) $25-38.

Milk glass bowl and stem with black foot. Morgantown Glass. Goblet. 6 1/4 inches. $20-26.

Footed Tumbler Belton pattern #9074 Morgantown Glass. Black bowl, green foot. See same form in crystal crackle bowl and black foot later in this chapter. (Snyder 1998) $80-100.

Black glass with cased bowl. Inside of all bowls is milk glass. Paper label on foot reads "Made In Italy." Form illustrated in ads for imports by Colony in the 1950-70s in black and white as well as other color combinations. Wine 4 1/4 inches, parfait 5 inches, sherbet 3 3/4 inches and goblet 6 inches. Any form. $10-19.

Lily pad feet circa 1930s. This form is known to have been produced in black and crystal combinations by Seneca Glass, McBride Glass, Fry Glass, and Louie/Weston Glass. Fry and Seneca should be lead crystal while Weston area production and McBride are soda lime glass. Some claim a distinction can be made based on the manner of crimp on the foot as imparted by the tool used. This theory has not proven sustainable when we looked at diverse samples. Some specific other colors and additional decorations do allow attribution to manufacturer. At present we believe the black lily pad pieces remain largely indistinguishable. L to R: measurements are heights: Sherbet 2 1/4 inches, water tumbler 5 1/2 inches, whiskey 2 1/4 inches, juice 3 3/4 inches and iced tea 5 inches. Any form. $8-16.

Princess gray and black. #6123 Fostoria Glass 1972-1980. Goblet 6 1/2 inches. (Long & Seate 2000) $20-28.

Blown hobnail bowl on black base. Unattributed. Sherbet 2 1/2 inches. $6-14.

Chalice pattern #6059 by Fostoria Glass. Sherbet 3 1/2 inches, cordial 2 3/4 inches. 1955-1959. Note the stem consists of three black wraps around a marble/sphere of crystal. Sherbet, $26-32; cordial, $38-45.

Lily pad footed juice, $8-12, and tumbler, $10-16. Fry Glass. See discussion with black and crystal forms. The only known manufacturer of this color combination in Lily pad pieces is Fry.

Blossom-Anisette pattern #X1072 Mikasa Crystal. 1980-1981 only. Goblet 9 1/4 inches. Note the black stem wraps up onto the optic patterned bowl. (www.replacements.com) $28-34.

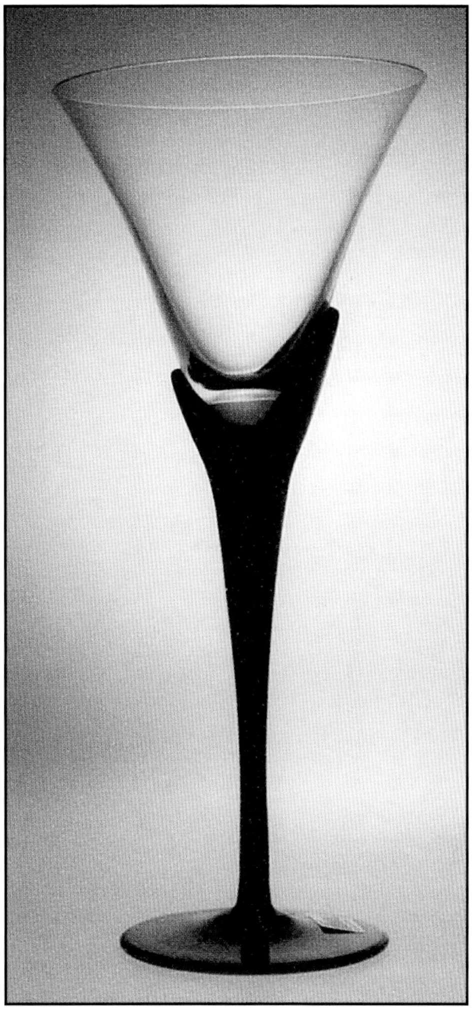

Vogue line #TS201 Mikasa Crystal. 1987-2003. Goblet 9 3/4 inches. Another of the 1980-'90s styled black stems that wrap onto the bowls. $32-42.

Tulip in black. Mikasa Crystal circa 1987. Goblet 9 1/4 inches. (www.replacements.com) $30-38.

Lotus in black. #6144 Fostoria Glass. 1980-1982 only. Goblet 8 1/4 inches. $28-38. Champagne flute 8 3/4 inches. $30-38. (Long & Seate 2003)

Gazebo #6126 Fostoria Glass black stem and foot with crystal optic bowl. 1980-1982. Goblet 7 inches. (Long & Seate 2003) $20-26.

Tulipe black by Villeroy Boch Crystal. 1983-1987. Goblet 7 1/4 inches, $25-36, and iced tea 6 1/2 inches, $26-34. Note the black petal-like stem wraps up onto the crystal bowl. Paper labels remain on these bowls in this image noting the manufacturer.

Sutton Place. Lenox Crystal. 1980-1982. Goblet 7 1/2 inches. (www.replacements.com) $24-30.

Eloquence Onyx #6120 Fostoria Glass. 1970-1974. Goblet 7 inches. (Long & Seate 2003) $22-30.

Carnevale black by Sasaki Crystal. 1995-1999. Goblet 9 inches. (www.replacements.com) $10-16.

Royale line. Sasaki Crystal black foot and stem with twisted optic crystal bowl. Goblet 7 3/4 inches. (www.replacements.com) $30-36.

Glamour Onyx line. #6103 Fostoria Glass. 1969-1975. Goblet 7 inches. (Long and Seate 2003) $20-26.

Midnight Mood line. #17683 by Tiffin/U.S. Glass. Goblet 7 1/2 inches, black foot and stem, crystal straight optic bowl. (www.replacements.com) $24-30.

Sergio stem line. Sasaki Crystal. Circa 1991. Asymmetrical black stem and foot with crystal bowl. Goblet 8 3/4 inches. (www.replacements.com) $32-42.

Tucano line. Sasaki Crystal. Gold accent on black stem and foot with crystal bowl. Goblet 8 3/4 inches. 1991-1999. (www.replacements.com) $45-55.

Daniel Hechter. Marked on base in script by decal. Goblet 9 1/4 inches. Circa 1980s. Daniel Hechter was a division of Mikasa. (www.replacements.com) $12-20.

Seville black. Union Street Glass. 1996-2000. Crystal foot and bowl with black stem. Heavy optic patterned bowl. Wine 8 1/2 inches. $30-38.

Elysee by Mikasa Crystal. Wine 7 1/4 inches. (www.replacements.com) $12-16.

Onyx by Cristal D'Arques (France) circa 1990s. Goblet 7 inches. (Page & Frederiksen 1997) $10-14.

Silhouette line. #17707 Tiffin Crystal. Goblet 7 1/2 inches. (Page & Frederiksen 1994) $22-30.

Octime black by Arcoroc glass. Goblet 8 1/4 inches. (www.replacements.com) $12-16.

Game set footed tumbler Cambridge Glass. Club base, 5 inches. Circa 1930s. Each, $75-95.

Ringed and optic bowl sherbet with black foot. In the manner of several paste mold, hand blown West Virginia factories. Possibly McBride Glass or Weston area company. 4 3/4 inches. $8-12.

Ringed and optic bowl with black foot and gray cutting. The shape is found in several Weston glass catalogs, as shown possibly Louie Glass. Iced tea 6 inches. Circa 1930s. $10-14.

Crackle bowl, Belton pattern footed tumbler #9074. Morgantown Glass. 5 1/4 inches. See similar form earlier in this chapter with black bowl and green foot. (Gallagher 2001) $20-34.

Black foot pilsner – in the manner of several West Virginia paste mold, mouth blown companies circa 1920-1960. Definitive attribution not made. $8-14.

Sherbet. Central Glass. 4 inches tall. (Schmidt 2004) $18-22.

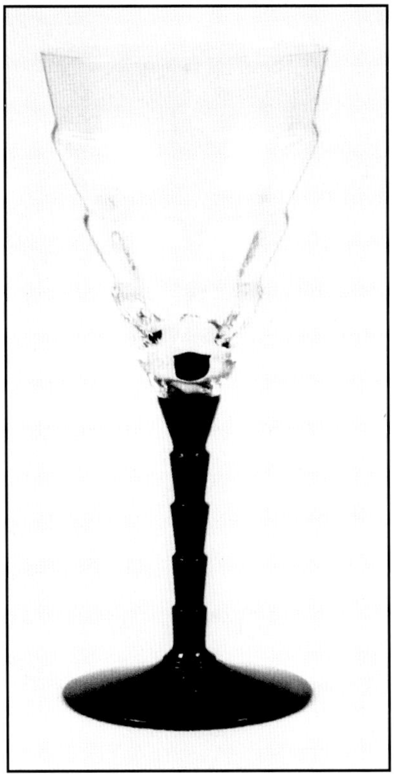

Moderne. #1446 Central Glass. Goblet 7 1/2 inches. Note this is the same bowl form used on the sherbet shown above. (Schmidt 2004) $32-46.

Central Glass, the same black base used with various crystal bowls. Left to right: Garland (original name) etching #1001 by Lotus Glass, later renamed McGuire etching by author, Weatherman. Central Glass blank, goblet 5 3/4 inches, $38-48; Harding etch #401 iced tea 6 1/4 inches, $90-115; line #1451 with "V" shaped bowl 5 3/4 inches, $16-22; line 1450 bell shape bowl 5 3/4 inches, $18-22. (Weatherman 1974 and Schmidt 2004)

Aurora #1066. Cambridge Glass. Circa 1931. The black stem is a faceted ball, which unfortunately does not show well in this image. Liquor cocktail 3 1/2 inches. (Cambridge Collectors 1976) As shown, $22-26.

Topaz bowl and black foot and stem, Sloan Glass. Circa 1930-1932. Note these stems are distinctively three sided and were a patented design. Sherbet 4 1/4 inches. $12-20.

Footed tumblers, black foot and crystal straight ribbed optic bowl. Shown here with gray leaf cutting. Circa 1930s Weston glass companies/ Louie Glass. Iced tea 6 inches and water 5 inches. Either, $8-12.

Footed tumbler black foot and crystal straight ribbed optic bowl. Identical form to that shown nearby with gray cutting. 3 inches. As shown, $4-8.

Saucer champagne or sherbet. Vision pattern #3008 Ebony. Morgantown Glass. 1971-1973. 5 1/2 inches. (Snyder 1998) $8-16.

Black goblet. Patriot pattern #3014. Morgantown Glass. 5 1/2 inches. (Snyder 1998) $12-16.

Paragon #7624. Morgantown Glass. Introduced 1929. Sherbet or saucer champagne 5 1/4 tall. (Gallagher 2001) $75-110.

Fashionable line #1974. Seneca Glass sherbet 4 1/2 inches, $12-14; goblet 6 1/4 inches, $14-16; iced tea 6 3/4 inches, $16-18. Circa 1971-1983. (Seneca catalogs)

Venture line. Lenox Crystal. Decorated with platinum banding. 1975-1982. Goblet 7 inches. (Page & Frederiksen 1997) $22-30.

Biscayne Black Fostoria Glass circa 1971-1973. Goblet 6 1/2 inches. (Long & Seate 2003) $16-24.

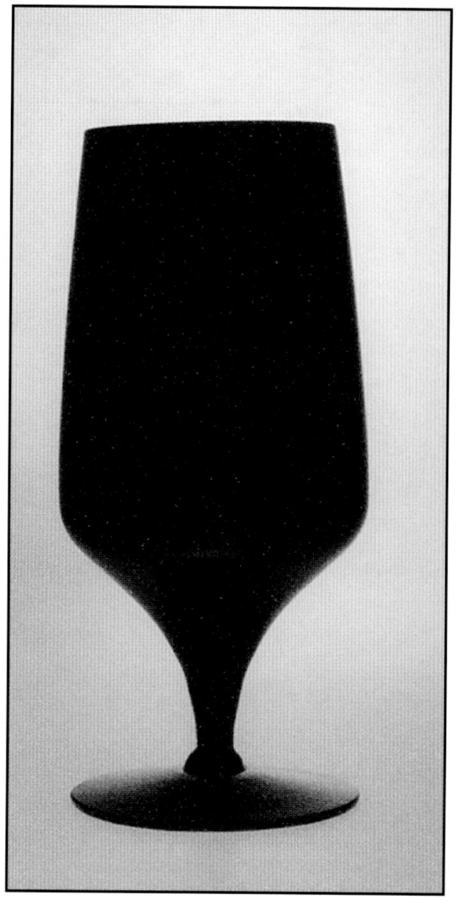

Vision line #3008. Morgantown Glass. Iced tea 6 1/2 inches. Circa 1971-1973 (Snyder 1998) $18-22.

Octagonal footed tumbler or vase #148. Beaumont Glass. Circa 1930s. $26-36.

Footed and handled tumbler with enamel red and black rings. Attributed to Huntington Tumbler. Circa 1930s. 5 inches. $12-20.

American Manor #17703. Tiffin Glass. Ebony. Sherbet or saucer champagne 5 3/4 inches. $12-14.

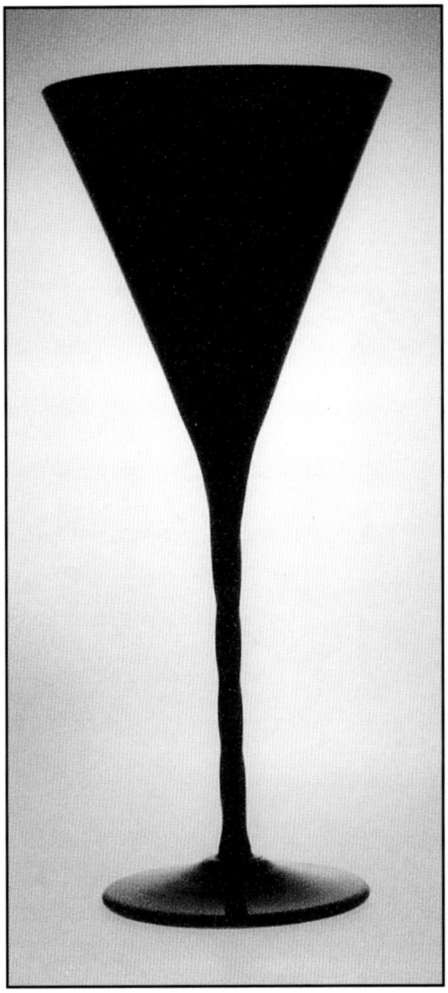

Braid black #SX725 by Mikasa Crystal. 1986-1987 only. Goblet 9 1/2 inches. (www.replacements.com) $20-28.

Midnight by Toscany Crystal. Goblet 7 inches. (www.replacements.com) $14-20.

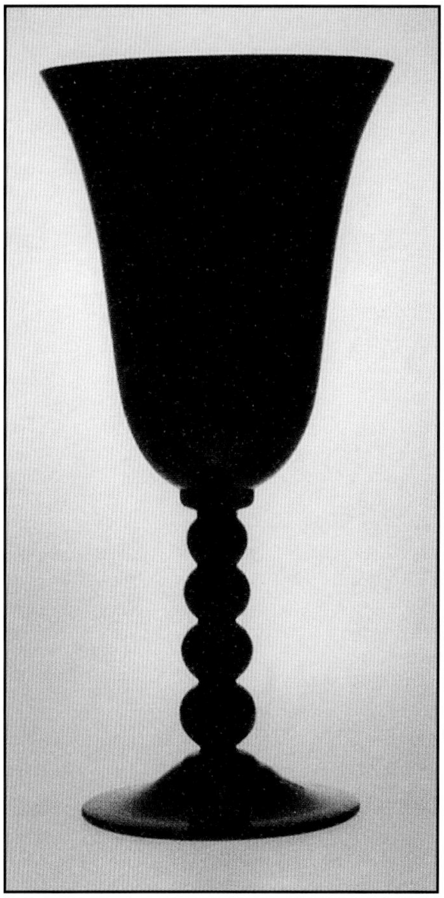

Candlewick black. #3400 Imperial Glass goblet 7 1/2 inches. Circa 1938-1945. (Garrison 2003) $160-185.

Jet Black by Reizart. Circa 1958. Iced tea 6 1/4 inches. (www.replacements.com) $20-26.

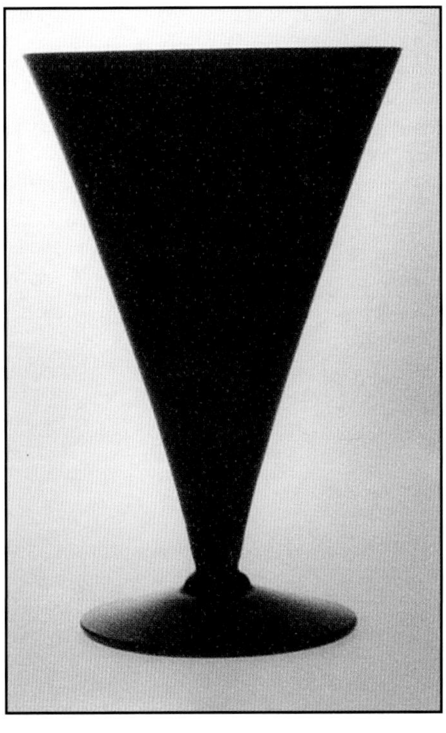

Impromptu by Lenox Crystal. Circa 1974-1979. Tumbler 5 5/8 inches, $10-16; goblet 7 inches, $14-18. Note: these are heavy, thick walled objects in the casual-textured manner of the 1970s. (Page & Frederiksen 1997)

Spotlight ebony by Noritake Crystal. 1974-1983. Iced tea 6 1/2 inches, $10-14; goblet 6 inches, $8-14. (Page & Frederiksen 1997)

Octagonal Black by Independence Crystal. Circa 1970. Iced tea 6 1/8 inches, $10-16; sherbet 4 inches, $10-14. (Page & Frederiksen 1997)

Perspective by Noritake Crystal. 1974-1983. Goblet 6 1/2 inches, $10-16; rock glass 4 inches, $12-16. (Page & Frederiksen 1997)

Sweet Swirl Black by Noritake Crystal. 1985-1994. Tumbler 5 1/8 inches, $8-12; goblet 7 1/2 inches, $10-16. (Page & Frederiksen 1997)

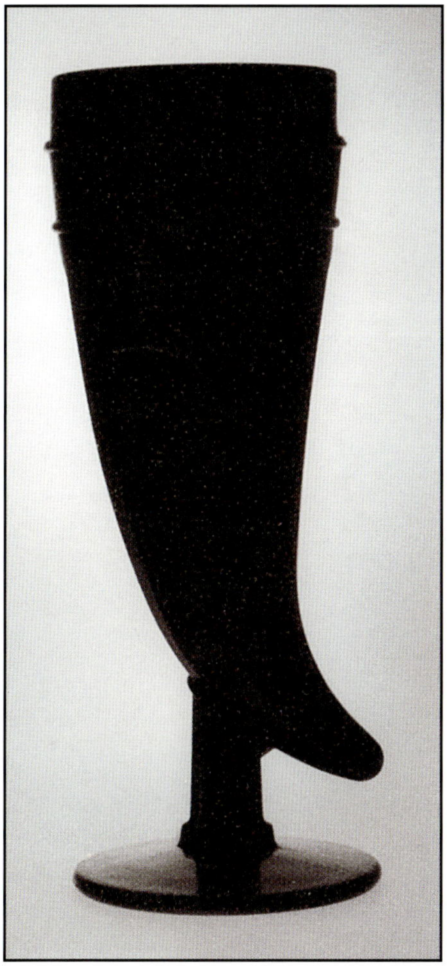

Left:
#10016 Hunter Horn mug, Tiara Glass. 8 inches. Conceived 1980 by Arthur Harshman, Tiara's Director of Design, and made for many years in crystal. Black was offered in 1982 only. This is machine made glass and not handmade glass. Thus, it is not a product of Viking Glass as has been repeated in several in print sources. $10-16.

Below:
Top Hat ice bucket, Blenko Glass. $80-100.

Decagon pattern ice bucket. Cambridge Glass, circa 1930s. 5 1/2 inches. $95-110.

Farber ware ice bucket. Farber ware is the manufacturer of the chrome stylized holder. Pieces often marked Farber. Glass inserts by several companies and this is likely to be a Cambridge Glass product. Morgantown is another likely glass supplier to Farber. Circa 1930s. $90-110.

Ice bucket, hand blown with clear applied handles. Beaumont Glass. Circa 1980s. 7 3/4 inches tall. $45-65.

Cowboy hat ice bucket. Blenko Glass. $75-90.

Right:
Lynward pattern black decanter with crystal stopper. Morgantown Glass. 10 1/2 inches. See the company history for period illustrations of this form and other Morgantown decanters. $150-200.

Hand blown martini pitcher. Unattributed. Crystal with black foot and handle. 8 3/4 inches. $22-32.

Left:
Hand blown pitcher. Crystal with vertical optic and handle. 8 3/4 inches. $18-28.

Right:
Morgantown decanter with crystal stopper and silver overlay. The form was named Little King by Weatherman. (Weatherman 1974) Note: please see the Morgantown company history later in this book for an additional image of this form and other decanters. As shown, $200-250.

Crystal decanter with black stopper. Steuben Glass. 10 inches. $450-600.

Tilt Ball Decanter, black with crystal stopper, Cambridge Glass, 7 inches tall. Circa 1930s. $100-125.

This long necked decanter is shown as an example of the black art glass coming into the U.S. from the 1950s until the present from Italy and numerous other European countries. An entire book could be dedicated to this source alone. We include this graceful, but very un-American, blown and elongated form here as a reminder of the millions of pieces of black glass imported to the U.S. and largely not addressed in this book project. 19 1/2 inches. $35-50.

Drink stirrers. Maker unknown. $4-10 each.

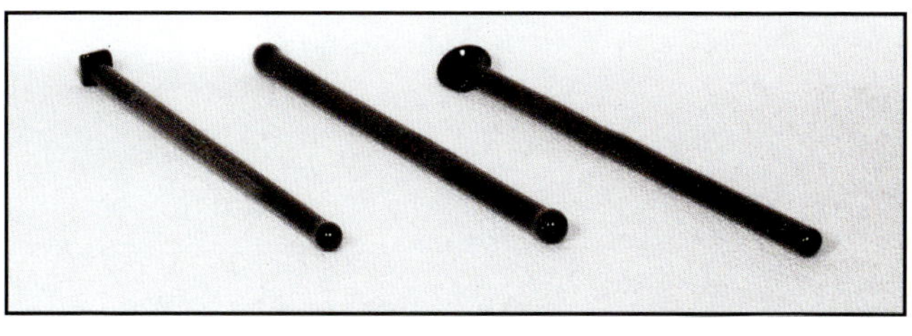

Chapter Eleven
Tableware, Plates, and Serving Pieces

Algues pattern dinner plate. Lalique (France). 11 inches. Satin finish, hand signed "Lalique" in script on reverse. $170-200.

Open edge square salad plate. Westmoreland Glass. 8 1/4 inches. Circa 1950s. $18-24.

Candlewick pattern bread plate. Dalzell-Viking Glass. 6 inches. Note that most black Candlewick will be manufactured by Dalzell-Viking from original Imperial Glass molds. $12-18.

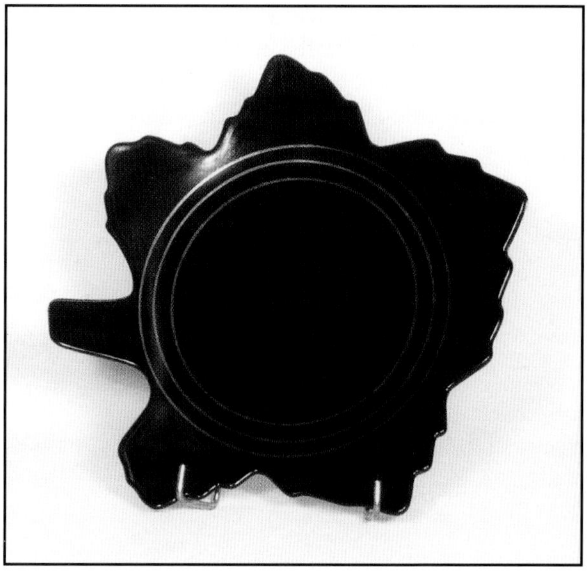

Leaf salad plate. Westmoreland Glass. 7 1/2 inches, with silver band decoration. $14-24.

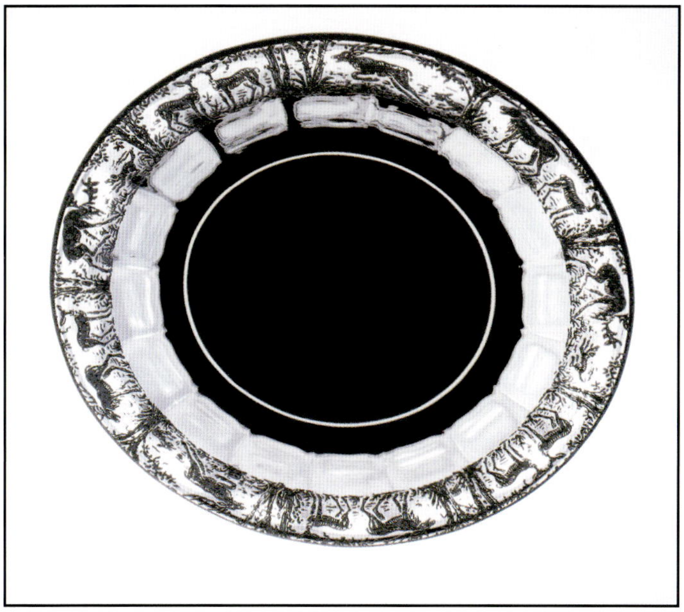

Black Forest etch, Paden City Glass line #210. Note: what appears as white panels where the rim of the plate falls in the center is actually the reflection of light from taking the photograph. The plate is all black. (Domitz 2004) $95-100.

Plate #34 with humming bird decoration. New Martinsville Glass. Circa 1930. Unknown decorator. $18-28.

Detail of humming bird decoration.

Satin glass plate. U.S. Glass. 9 inches. $14-16.

Charger or torte plate. Dalzell-Viking. Ground bottom. 14 inches. Classic Round pattern. $14-26.

Decagon pattern cracker plate. Cambridge Glass. Part of a cheese and cracker set. Note the well for the stemmed cheese dish in center. 10 1/2 inches. Plate only, $22-28.

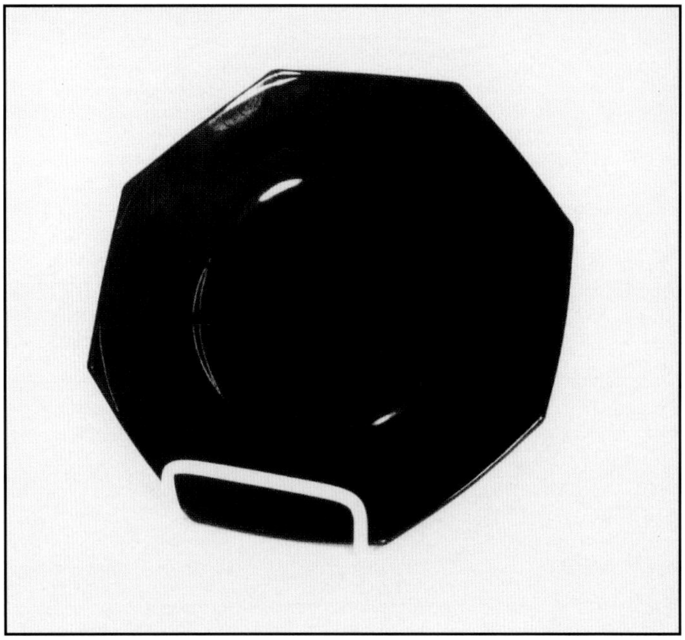

Octagonal salad or luncheon plate. Ground bottom. 7 3/4 inches. Unattributed. $6-8.

Scroll pattern. U.S. Glass. Left, salad or luncheon plate 7 1/2 inches and bread & butter (?) 5 inches. Some pieces in this pattern may bear the small mark of the USG embossed logo on the reverse and are slightly more desirable. Named by collectors. As shown, $6-10.

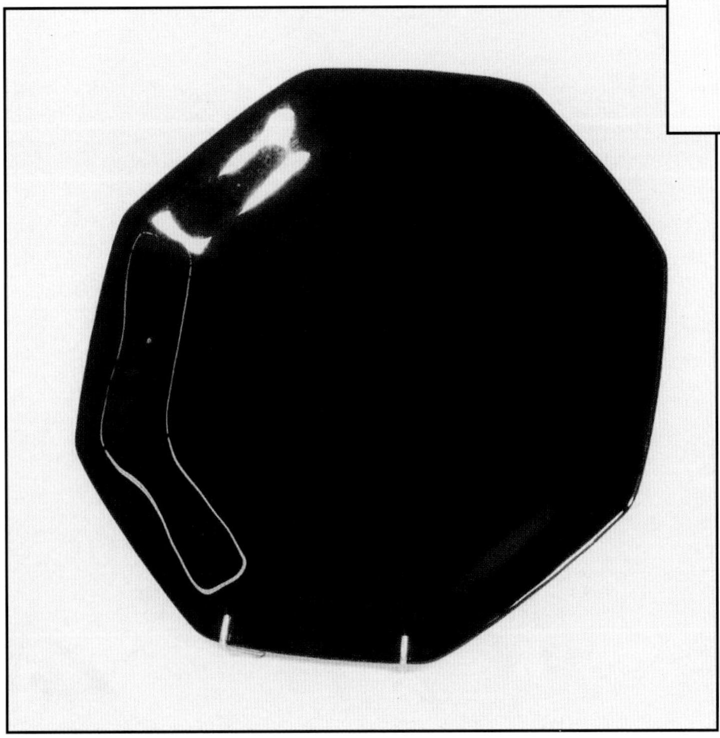

Octime pattern dinner plate. Arcoroc Glass. "Made in France" embossed on back. 10 3/4 inches. $5-8.

Octagonal plate with swag pattern. Possibly Smith Glass due to similar design elements? 7 1/2 inches. 1930s. $12-14.

Cake or torte plate reverse showing rayed bottom. 10 inches. There are many of these similar relatively plain patterns, each lacking details that allowed us to distinguish them. Unattributed. $8-14.

Right:
Mayfair pattern salad or luncheon plate #2419. Fostoria Glass. Manufactured 1930-1940. (Long & Seate 2003) $10-14.

Cloverleaf pattern. Hazel Atlas Glass. Sugar bowl, 3 inches height, $14-20; 8 inch plate, $16-22; cup and saucer, $20-28; sherbet, $16-22; and creamer, 3 inches height, $14-20.

Cake or torte plate reverse showing rayed bottom. 9 1/4 inches. Unattributed. $8-14.

Charade pattern luncheon or salad plate. Diamond Glass. 8 inches. $8-14.

Oriental silver deposit decoration. Rockwell Silver Co. on unattributed glass. Plate 10 1/2 inches, $95-125; cup and saucer, $55-80. Replacements Ltd. Museum Collection. Silver signed "Rockwell Sterling."

Line 1450 Central Glass plate, $10-16, and grapefruit bowl, $16-22; cup and saucer pair, $30-36. (Schmidt 2004)

English Hobnail pattern cup and saucer. Westmoreland Glass. Crystal cup with black glass saucer, 5 inches diameter (inverted to show pattern). $18-24 per set.

Octime pattern mug, $6-8; cup and saucer, $4-6. Arcoroc (France). Mug, 4 inches tall, marked "made in France" on reverse. Contemporary production.

Plain round tea cup. Unattributed. $4-6.

Octagonal cup and saucer, Beaumont Glass with silver decoration. $20-25 per set.

Humming bird decoration on octagonal line #34 New Martinsville Glass. See plate shown earlier in this chapter for more information. $24-34.

#10304 Cameo pattern covered oval butter dish. Indiana Glass for Tiara Exclusives. 9 inches. Circa 1977-1981. This pattern was originally issued by Indiana as Diamond Point in 1965. $16-24.

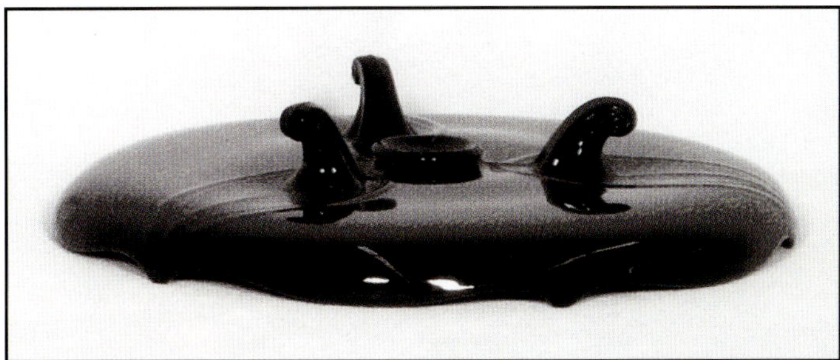

Three toed plate. Scalloped and rolled rim. 10 inches. We suspect Smith Glass based on similar design and style elements but unattributed. $14-18.

Fuchsia decoration. Lotus Glass decorators. Mayonnaise and under plate. Gold decoration #905 with plate etching. (Six 2005) Piece type, $38-48.

Imitation cut starburst pattern two handled bowl. Imperial Glass. 8 1/2 inches. $28-38.

Salt dish. Unattributed. Panel pattern. 2 inches tall. Note: this piece may be a child's sherbet or individual almond dish. $24-36.

Sugar with crystal handle #3400/98, $22-28 on left. Jug with ice lip and crystal handle #3400/38 80 oz., $115-130. Cambridge Glass. Appearing in Cambridge catalog reprint 1933. (Cambridge Collectors 1976)

Crinkle pattern. Morgantown Glass. Sherbet 3 inches high. While this was a popular and relatively long produced pattern, it is uncommon in black. $20-25.

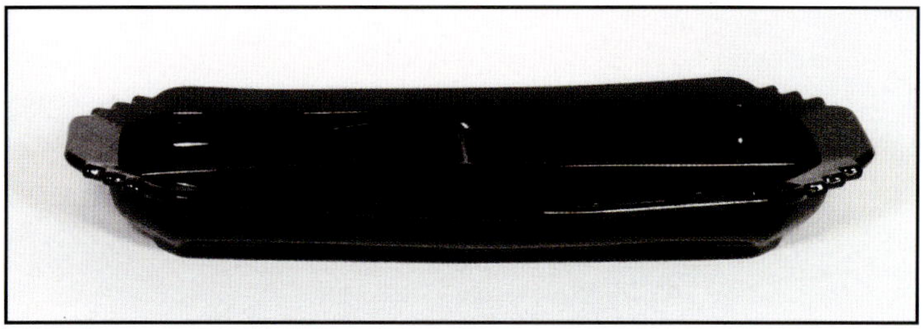

Divided relish line #1450. Central Glass. 3 part. 12 1/4 inches. (Schmidt 2004) $28-38.

Tumbler and pitcher. Cameo or Diamond Point pattern (same pattern, both names used). Tiara Exclusives/ Indiana Glass. 1977-1981. Tumbler, $6-10; pitcher, $14-24.

Apple Blossom etch #3400/647. Cambridge Glass. 5 part relish. The decoration is actually sterling silver, contrary to how it photographed, and is decoration D-970 Apple Blossom. (Cambridge Collectors 1976) $150-195.

Cherry Thumbprint pattern. Mosser Glass. Tumbler 4 inches on left, $10-16; jug 7 1/2 inches, $38-42. Circa 1991.

Contemporary jug. 8 1/4 inches. This jug/pitcher came to us marked "Made in China." $6-10.

Jug #935. Cambridge Glass. 64 oz., 8 1/4 inches. Appears in 1931 catalog. This flattened jug/pitcher was also made by the Weston area glass companies in depression green and pink, crystal, ruby and cobalt, and in black, but the black Weston forms appear with crystal and often reeded handles. As shown, with black handle, seems to be unique to Cambridge. (Cambridge Collectors 1976) $100-125.

Victorian era mugs. Atterbury Glass. Left: 2 1/2 inches, $28-36; right: 3 1/4 inches, $28-36. Circa 1880s.

Contemporary Irish Coffee styled footed mug. Probable manufacturer Indiana Glass. 5 1/2 inches. $4-8.

Contemporary coffee mug. 3 1/2 inches. "Jack Daniels Old No. 7 Brand Old Time Tennessee Whiskey" decoration. Unattributed. $4-8.

Tumbler, marked Hazel Atlas on base. 3 3/4 inches tall. Circa 1930s. $10-14.

Close up of base, Hazel-Atlas tumbler.

Cream pitcher, pressed glass. 4 1/2 inches. Unable to determine age. Unattributed. No value established.

Celery dish with handle #681. L.E. Smith Glass. 9 1/2 inches. Circa 1930. (Weatherman 1974) $24-36.

Footed cream and sugar #2350. Pioneer pattern. Fostoria Glass. 1924-1943. (Long & Seate 1999) Pair, $40-54.

Cream and sugar colonial pattern. Similar to Imperial and others, no specific attribution made. 2 1/2 inches. Pair, $30-48.

Golden Ray cream and sugar with brass metal bases attached. Unattributed as to the glass producer, but the name Golden Ray was registered to G.H. French Co. This style was popular in the 1950s. See the Kappenman black glass web site noted in bibliography and in the introduction chapter for more information. Set, $20-45.

Sugar #34. New Martinsville Glass. Circa 1930s. (Weatherman 1974). As shown, $16-20.

Sugar and cream. Scroll pattern. U.S. Glass. Circa 1920-30s. Note sugar is inverted to show pattern on base. Other pieces in this table set are octagonal. Pair, $28-38.

Sugar and cream Ovide with white enamel decoration. Hazel-Atlas Glass. 3 1/2 inches. Pair, $18-26.

Sugar and cream Ovide, Hazel-Atlas Glass. Undecorated, shown nearby with decoration. 3 1/2 inches. 1930s. $20-26.

Sugar and creamer, four toed with panel and sculpted rim. #125 line Indiana Glass. 3 3/4 inches. (Measel and Wiggins 1998) Pair, $22-38.

Sugar and cream, Molly pattern, footed with silver decorations. Imperial Glass. Circa 1930s. 4 inches. Note two styles of cream and sugar were produced for the Molly line. One sits flat and the other, shown here, has a footed base. Pair, $26-40.

Cream and sugar #148. Beaumont Glass. Hexagonal with silver decoration. 4 inches. 1930s. Pair, $22-38.

Sugar, cream, and tray #1450. Central Glass. (Schmidt 2004). As shown, 3 pieces set, $80-100.

Cream and sugar, Deerwood etched decoration. U.S. Glass. Gold encrusted. 2 3/4 inches. Circa 1930s. Pair, $275-300.

Cream #3400/68. Cambridge Glass. Circa 1930s. 3 inches. (Cambridge Collectors 1976) As shown, $18-28.

Sugar with lid and cream. #5029 L.E. Smith Glass. Circa 1930s. (Weatherman 1974) Pair, $35-50.

Cream. Pillar Flute pattern #682. Imperial Glass. 3 1/4 inches. (Weatherman 1974) $16-24.

Sugar and cream. Mt. Pleasant pattern. L.E. Smith Glass. Circa 1930s. 2 3/4 inches. (Weatherman 1974) Pair, $28-38.

Sugar and cream. Decagon line Cambridge Glass. Etched decoration #408. Circa 1930s. (Cambridge Collectors 1976) Pair, $40-54.

Cream. Diamond Block #330. Imperial Glass. $14-18.

Cream. Modernistic pattern #33. New Martinsville Glass. With uncommon enamel decoration. Design patented 1928. (Six 2002) As shown, $38-48.

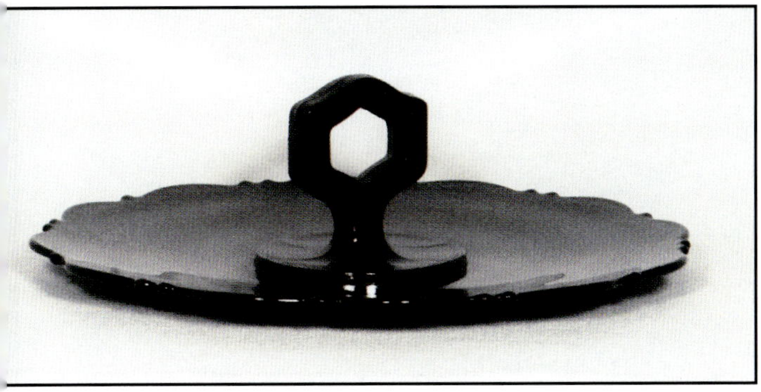

Sandwich tray. #505 Mt. Pleasant pattern, L.E. Smith Glass. 10 inches. Circa 1930s. (Weatherman 1974) $18-26.

Beverage tray. #381 L.E. Smith Glass. 6 1/2 inches. Circa 1930s. Literature at times calls this a "cordial tray." (Weatherman 1974) $12-16.

Tumbler tray with cup wells for six glasses. #181 L.E. Smith Glass. Circa 1930s. (Weatherman 1974) $14-20.

Sandwich tray with center handle and gold decoration. 10 inches. Unattributed. $16-24.

Cake plate, handled. #320 U.S. Glass. 10 inches. Gold band decoration. Satin. $24-36.

Sandwich tray with center handle. #148. Beaumont Glass. Octagonal. 10 1/2 inches. $18-28.

Sandwich tray with center handle. #725 Molly pattern, Imperial Glass. Circa 1930s. (Weatherman 1974) $22-28.

Handled bon bon. Casual line, Canton Glass. Circa 1951. Amber and black combination. (*All About Glass*, Vol. 1 No. 1.) $28-38.

Sandwich tray with center handle. #870 Cambridge Glass. 10 1/2 inches. Circa 1930s. (Cambridge catalog #27.) $32-42.

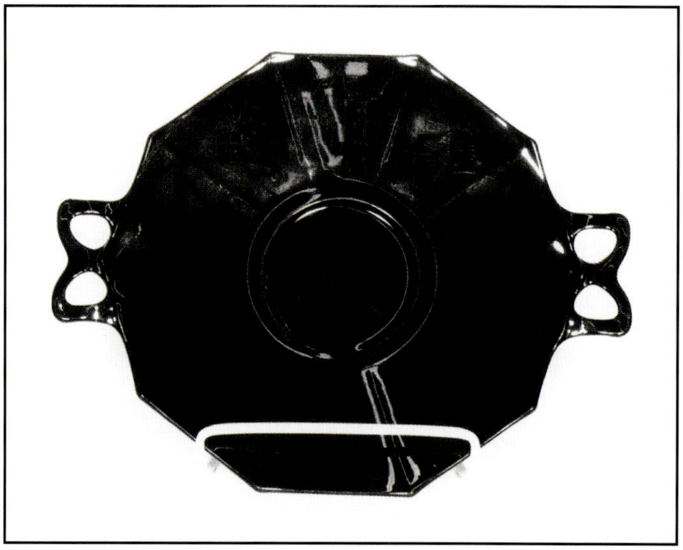

Cake plate, two open handles. #2375 Fairfax pattern, Fostoria Glass. 1930-1935. (Long & Seate 1999) $40-56.

Lunch tray with center handle. #2287 with decoration #57 Waveland, Fostoria Glass. Circa 1926. (Long & Seate 1999) $65-85.

Cake plate, two open handles. #210 Regina, Paden City Glass. Etched decoration Black Forest. 11 1/4 inches. Circa 1930. (Domitz 2004) $150-175.

Cake plate, two open handles. Line #411 Paden City Glass. Etched decoration Ardith. 10 inches. Circa 1930s. (Domitz 2004) $55-65.

Cake plate, square, two open handles. L.E. Smith Glass. Circa 1930s. 10 inches. (Weatherman 1974) $14-24.

Cake plate, round, two open handles. L.E. Smith Glass. (Weatherman 1974) 10 inches, $14-24; and right, 9 inches, $12-20.

Plate, salad. #200 L.E. Smith Glass with silver decoration. 8 1/2 inches. Note: literature calls this a salad plate. We believe it was intended to be used under a salad bowl, not as an individual serving plate. (Weatherman 1974) As shown, $18-22.

Plate, 8 inch, $6-10, and plate, two open handles, 10 inches, $14-24. L.E. Smith Glass. Circa 1930s. Note: 8 inch plate imaged showing back to show the hourglass-like designs appearing four times on the reverse of some, but not all, pieces called Mt. Pleasant pattern.

Molly pattern. #725 Imperial Glass. Circa 1930s. Open handled with gold decoration. Left to right: 7 inch mayonnaise underplate, six sided, $8-14; 10 inch cake plate, 10 sided, $12-18; 5 1/2 inches mayonnaise bowl, $10-16.

Two handled bowl #400/62 EC. Candlewick pattern. Imperial Glass. 8 1/2 inches. Circa 1937. (Garrison 2003) $200-240.

Punch bowl with silver ring decoration. Possibly Louie Glass. 7 inches high. A similar punch bowl is shown in Louie catalogs of the 1930s and was small in size, as is this. Marketed with under plate and small roly poly tumblers of the same form. As shown, $40-70. Complete set, $100-120.

Vegetable bowl, two part divided with lid. Paden City Glass with decoration by Lotus Decorating. 12 inches. (Domitz 2004) $150 180.

Dolphin, two handled bon bon plate. #1621-1703 Fenton Art Glass. 6 1/2 inches. (Whitmyer 2003) $20-28.

Bon bon. #81 L.E. Smith Glass. Note: the depressions in the rim make this easily confused for an ashtray. Smith literature calls it a bon bon. Circa 1930. $14-20.

Square Black Glass Tableware Patterns

This brilliant chart was created by Beverly Kappenman, whose website on black glass, http://members.aol.com/blkapp/index.html, we strongly endorse. WVMG acknowledges her work, the idea of creating a key by shape, and her willingness to allow us to use the data and format she devised.

Manufacturer's Name		Line#	Pattern Name	#Corner Points	#Side Points
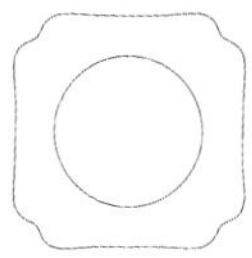	Cambridge	#3400	3400	2	0
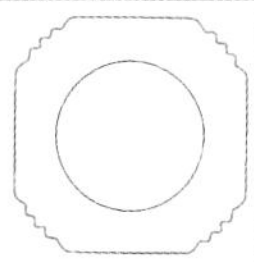	Central	#1450	1450	5	0
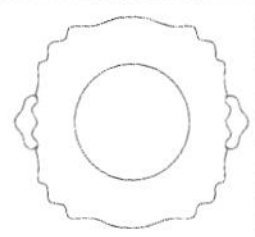	Central/Imperial	#-/148	Central/Imperial	3	1 (no plates w/ out handles known)
not pictured	Cooperative Flint	unknown	none	2.5	2
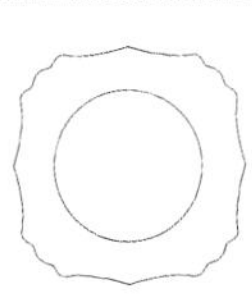	Diamond	#99	Charade	3	1
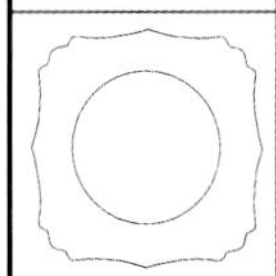	Fenton	#1639	Elizabeth aka Jay Cee	3	1
not pictured	Fostoria	blank #4020	4020	0	0 (no plates)
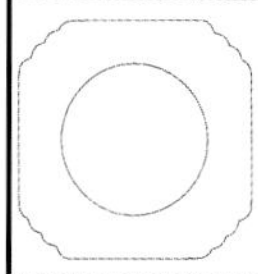	Fostoria	#2419	Mayfair	4	0
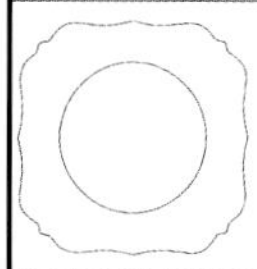	Fry ?	#?	Square	2.5	1
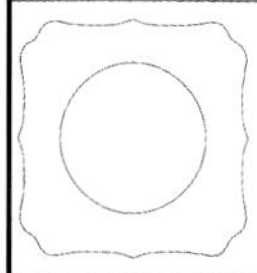	Imperial	#760	Hazen	2	1
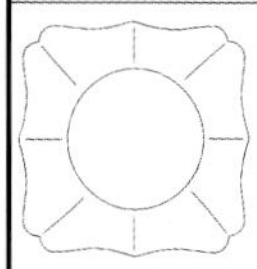	Liberty	#?	Truman	2	1
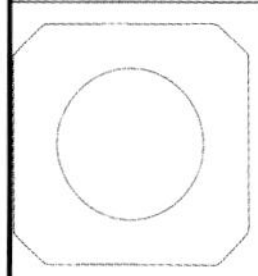	McKee	#?	Clico	0	0
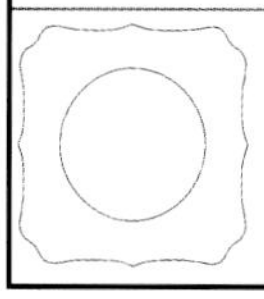	Morgantown	#1519/ #1520	Square	2	1
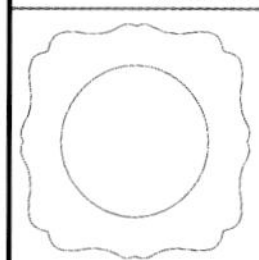	New Martinsville	#35	Fancy Square	2	2
not pictured	New Martinsville	#36	Square	4	3
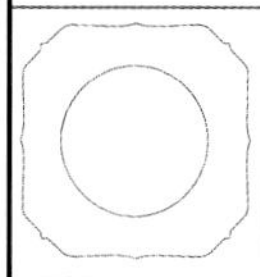	Paden City	#411	Mrs. B	3	1
not pictured	Paden City	#412	Crow's Foot Square	0	1
not pictured	L. E. Smith	#200 Series	none known	3	0
not pictured	L. E. Smith	#400 Series	Do-Si-Do	3	2
not pictured	US Glass/Tiffin	5800	Square	2	1
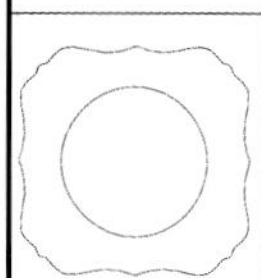	Unknown #1	–	–	2.5	1
not pictured	Unknown #2	–	–	3	1
not pictured	Unknown #3	–	–	5	0
not pictured	Westmoreland	1929-1	none	3	1

Chapter Twelve
Vases

Hobnail Vase #880. Kanawha Glass. Circa 1970s. Because of the swung form and hobnails, this is often called a Fenton product. The hobnails of Kanawha and general form are very different from the Fenton hobnail swung vases upon scrutiny. Swung vase so as per above note height varies. As shown 9 1/2 inches. (Kanawha Glass catalog.) $12-16.

Bud Vase #1140. Viking Glass. 10 1/2 inches. 1977-1978. (Viking catalog 1977 and 1984-85 catalog.) $10-14.

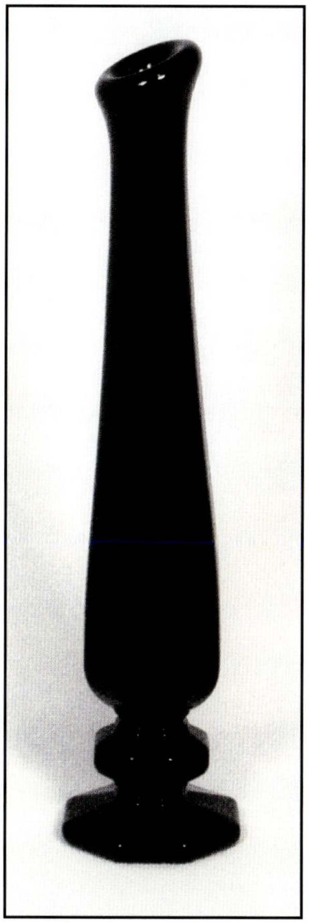

Flair Vase #800. Viking Glass. 9 1/2 inches. Shape was introduced by New Martinsville Glass in the 1940s as Christina in clear. (Six 2003) $12-18.

Bud Vase #9056. Fenton Art Glass. 8 1/2 inches, size varies. The vase is created by swinging it in an arc while the glass is still hot. This creates variances in height. $10-16.

Thumbprint pattern bud vases. Fenton Art Glass. Left: #4455BK, 11 1/2 inches, 1968-1976, $14-18; right: #4453BK. 16 inches, 1968-1976, (Whitmyer 1999) $16-22.

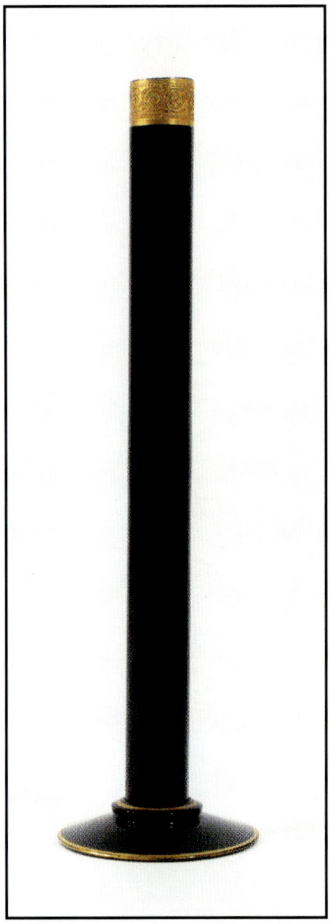

Stick Vase with gold encrusted decoration. 10 inches. This form causes considerable confusion in glass research as they are attributed to several companies. A difference that will in time allow us to sort some of them is that many have a button where the cylindrical vase meets the hand-set base/foot. Others do not. This example lacks that obvious "button" of glass where the two pieces were joined when being formed, and is the clue we are suggesting. We are not yet certain what that clue means. Possible manufacturers for this button-included form may be Maryland Glass, Central Glass, the Weston area companies, and certainly others. As shown, $18-28.

Left:
Thumbprint pattern vase 34454 BK. Fenton Art Glass. 1968-1975. (Whitmyer 1999) $16-20.

Right:
Stick vase with flared top. This vase continues the discussion begun above. Note that there is clearly no button where this cylinder joins the foot. And the illustrations in several Cambridge Glass catalogs and vases in colors unique to Cambridge feature no button. Thus this form can be attributed to Cambridge. It is uncertain at present if others used the same process to join the vase body and foot, or if this "button-less" stick vase is unique to Cambridge. As shown with enamel decoration, $18-22.

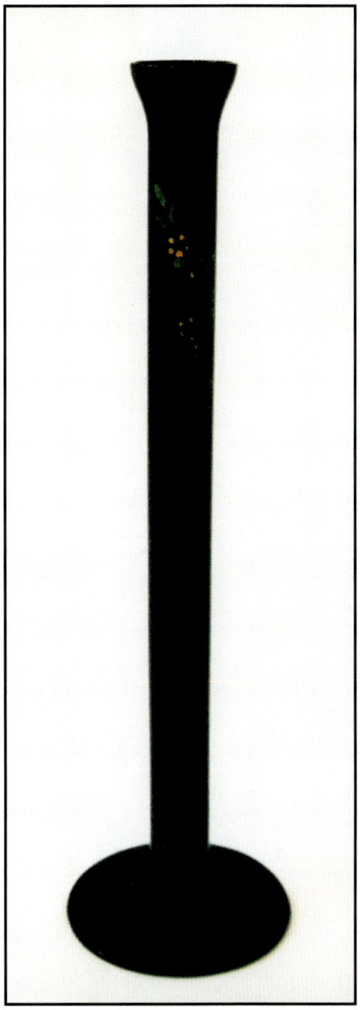

Stick vase. See discussion appearing with others of same form. This example has been attributed to Maryland Glass. That discussion remains open as per above. Note this is a buttoned form. As shown with enamel, $24-28.

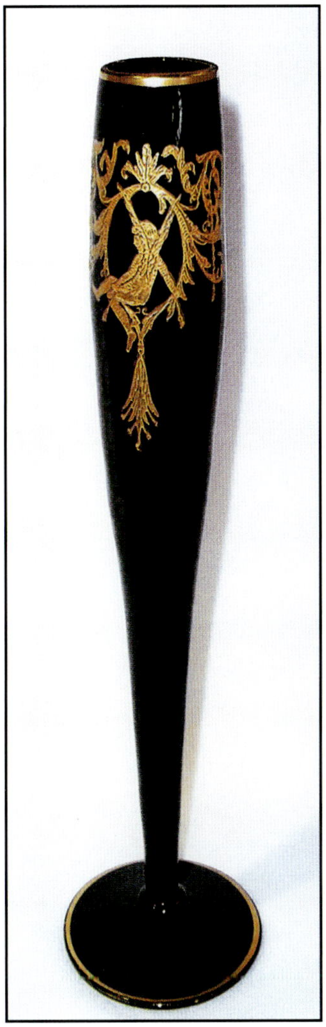

Cigar form vases with enamel decoration. All bearing Central Glass original manufacturers labels on base. All 10 inches. Note consistency of form and button where vase and foot are joined. Basket, $30-48; blue floral and swag, $45-65; peacock, $48-70.

Left:
Cigar shaped vase #1320/168. Central Glass with Morgan etch #412. 10 inches. Note: while this form is common and unlikely to be attributed without decoration, this etching is well documented and eagerly sought by collectors. (Schmidt 2004) $400-600.

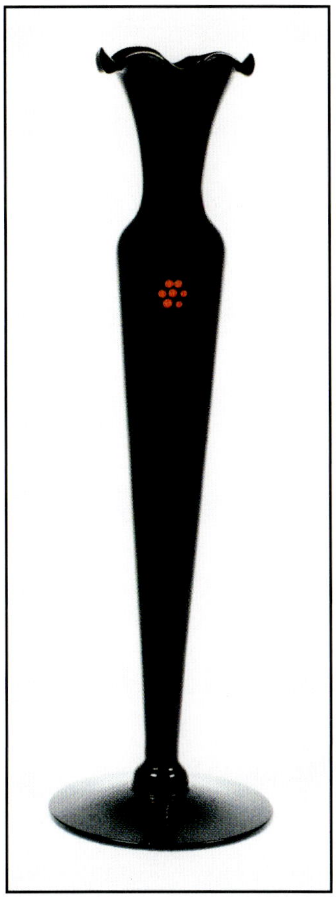

Left:
Crimp top footed bud vase. 10 inches. Circa 1930s. This decoration of red enamel dots forming a floral pattern and encircled by two gold bands which appear very faintly in the image, has long been found in the Weston area and is commonly attributed to the area but that is not yet a certainty. Again, these similarly-shaped small vases require some scholarship to establish better attributions. Who will accept this challenge? As shown, $10-14.

Right:
Footed, crimped top bud vase. Louie Glass. Note the similarity to the decorated similar forms. Circa 1930s. There are observable differences in the location of the waist of these vases when observed in profile, etc. The form, crimp, and manner of foot setting should be clues, once we learn how to read them, to ascertain the manufacturer. 10 inches. $10-14.

Decorated, crimped, and footed bud vases. This form is one of the most common made in American handmade glass houses. It was produced in crystal by as many as 4-6 companies well into the 1990s, and probably had similar production diversity in black in the early twentieth century. While both of these came to the book project with Maryland Glass attributions, we note what appear as differences in the crimps and in the manner of setting the foot on the two. For these forms we repeat our call for more research. As shown, either vase, $12-20.

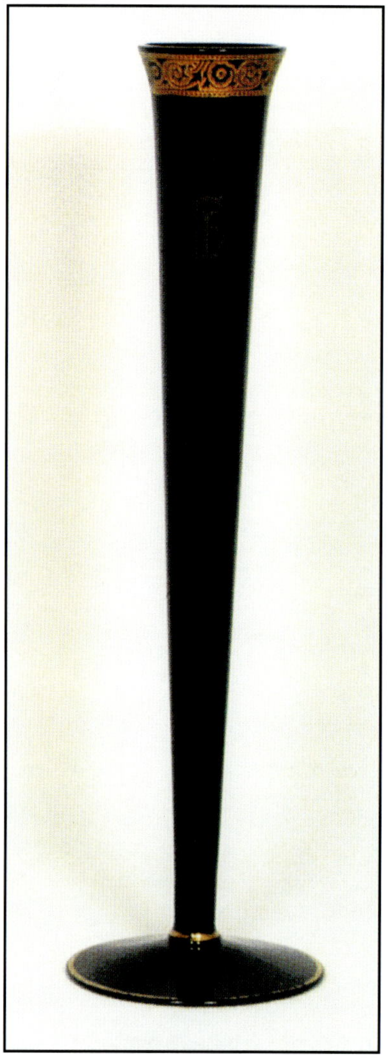

Vase. Minton Band gold encrustation decoration with gold encrusted monogram. 10 inches. Obviously an elegant, costly product in its day. Minton band was a decoration specifically made to match Minton China and a major producer of the decoration was Lotus Glass. The manufacturer of the glass and decorator remain unknown for this object. As shown, $12-20.

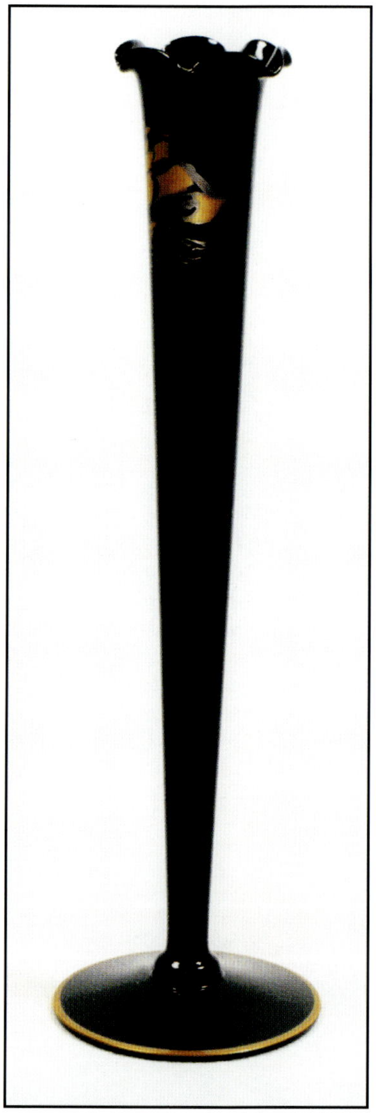

Gold sailing ship decoration crimped top, trumpet bud vase. Probably attributed to U.S. Glass due to decoration. 10 inches. Note the crimp at the rim and compare to others. There are distinctive crimping tools unique to some manufacturers. This merits study as well. As shown, $45-55.

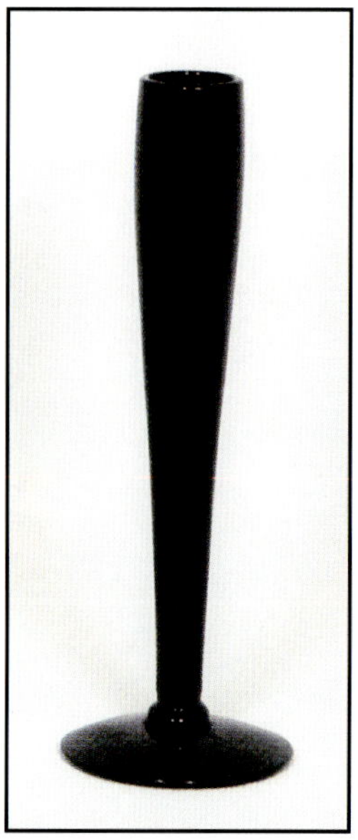

Cigar form bud vase. Louie Glass. Circa 1930s. Shown is a less common short size of 6 inches, $10-16.

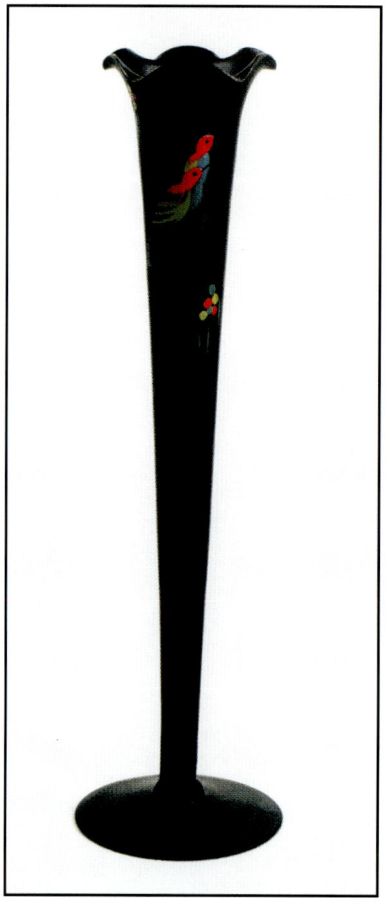

Trumpet vase #9723 U.S. Glass. Hand enameled bird decoration. 10 inches. Again it is the known and documented decoration that allows attribution of a form that many, many glasshouses made. As shown, $30-45.

Gold sponge-ware like decorations. The image of these two vases came to this book project with attributions of being from the Maryland Glass Co. We do not know to the contrary, and accept this pending additional research. As shown left, $14-24; right, $16-24.

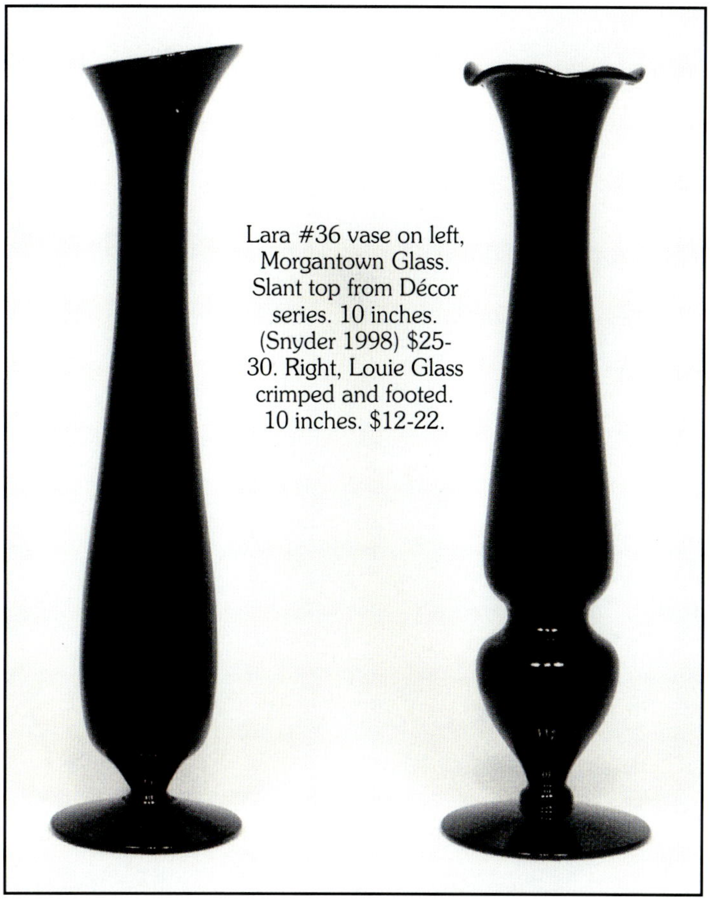

Lara #36 vase on left, Morgantown Glass. Slant top from Décor series. 10 inches. (Snyder 1998) $25-30. Right, Louie Glass crimped and footed. 10 inches. $12-22.

Vase, footed and stepped waist. In the style of the Weston glass factories but no attribution affirmed. 10 1/2 inches. $18-28.

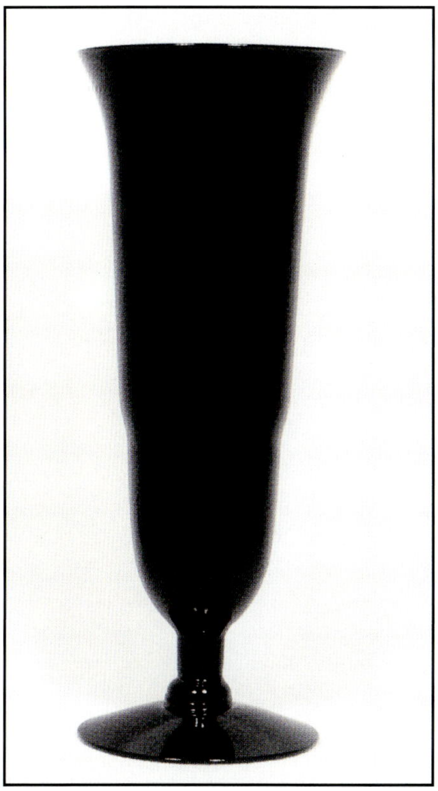

Footed Gold band vase. 7 1/2 inches. Ground top. Dragon-sphinx decoration. One possibility is that with the ground and not fire polished rim and the mythological beasts this may be of English/European origin. A vase with similar decoration has been attributed to Maryland Glass, but that vase has a fire polished rim opening. There is the possibility of this being a piece from a decorating shop using variously sourced glass blanks. Otherwise it remains Unattributed. $30-42.

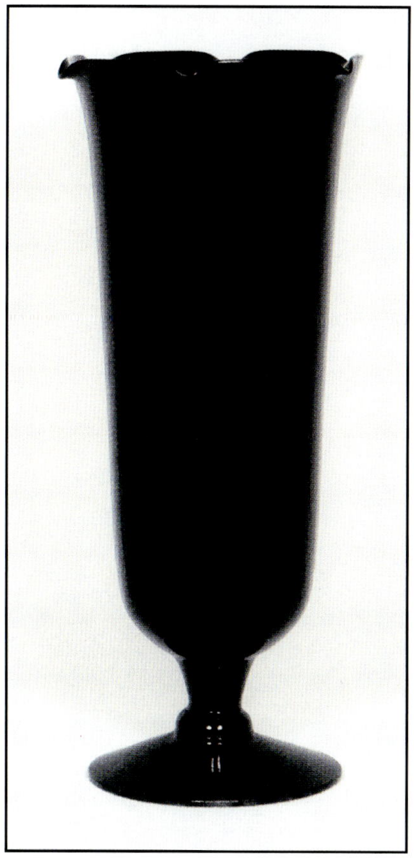

Vase, crimped top and footed. In the style of the Weston glass factories but no attribution affirmed. 10 inches. $18-28.

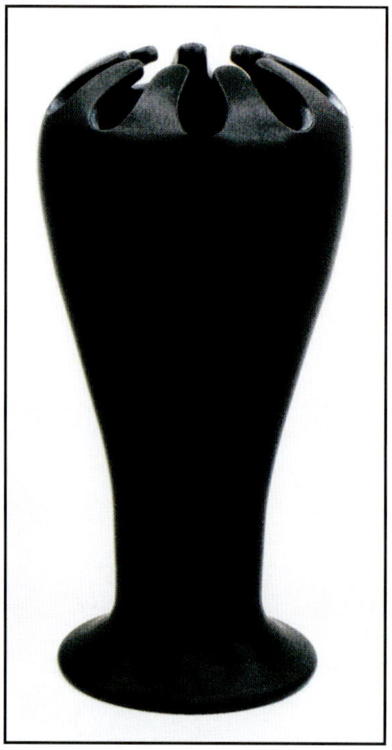

Left:
Crimped top vase #7348. Fenton Art Glass. 6 inches. 1985-2001. (Walk 2003) $18-28.

Right:
Flower arranger #16261. U.S. Glass. 8 1/2 inches. Circa 1927. Satin. Note a 12 inch version with a footed stem was also produced. $45-60.

Cupped Dahlia Vase #15151. U.S. Glass. Circa 1920s. 10 inches. Glossy finish. This vase came with decorations and glossy finish, an example of each also shown here. As shown, $42-48.

Fluted top, footed vase. 8 1/2 inches. Rough pontil, thick blown glass. Hand enameled floral decoration. Note the rough pontil scar on the base. Form and decoration together suggest a style at times generically called "Bristol," making a loose attribution to a blown style from the Bristol area of England. Much of this styled glass is late nineteenth and early twentieth century European glass. The piece has many indicators that suggest it is not American in origin. Circa 1900. $35-55.

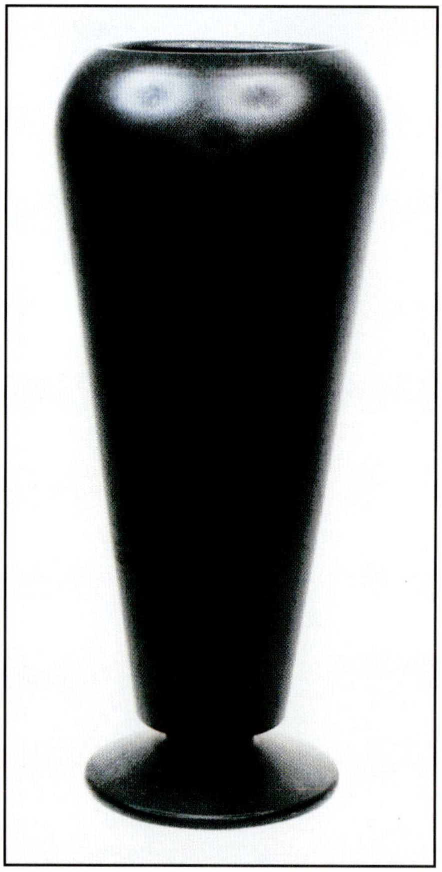

Cupped Dahlia vase #15151. U.S. Glass. Circa 1920s. 10 - 10 1/2 inches, cupping causes height variances. Satin. $48-56.

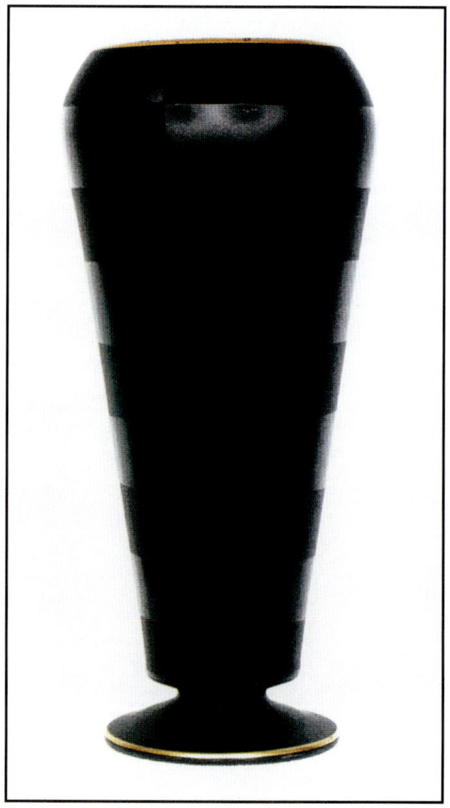

Cupped Dahlia Vase #15151. U.S. Glass with Echec Decoration. 10 inches. $80-100.

Gothic Garden etched decoration on vase #411, Paden City Glass. 4 3/4 inches. (Domitz 2004) $85-100.

Trumpet vase with rolled edge. 9 1/4 inches tall. Unattributed. $35-45.

Three footed vase #12. Duncan & Miller Glass. 1930s. $160-220.

Lily vase #151, 8 inches on left, $30-36, and Flared Dahlia Vase #15151, 8 inches on right, $36-40. Both satin and U.S. Glass.

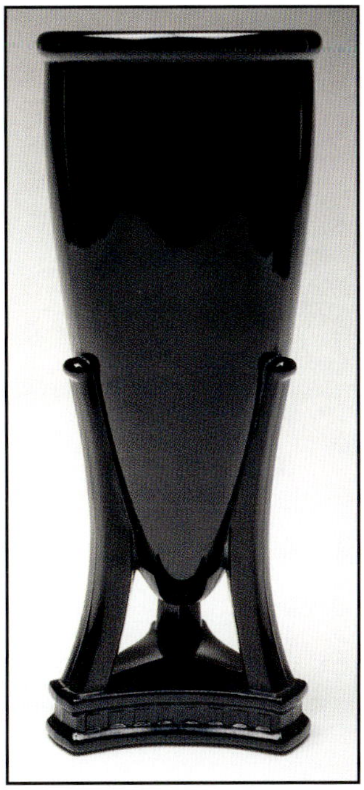

Tripod Vase. #2454 Fostoria Glass. 8 inches. 1932-1938. (Long & Seate 2000) $85-95.

Tripod Vase. #2454 Fostoria Glass. Unknown gold encrusted decoration. 8 inches. 1932-1938. (Long & Seate 2000) $100-150.

Janice pattern flared three toed #4527 vase. Viking/New Martinsville Glass. $120-140.

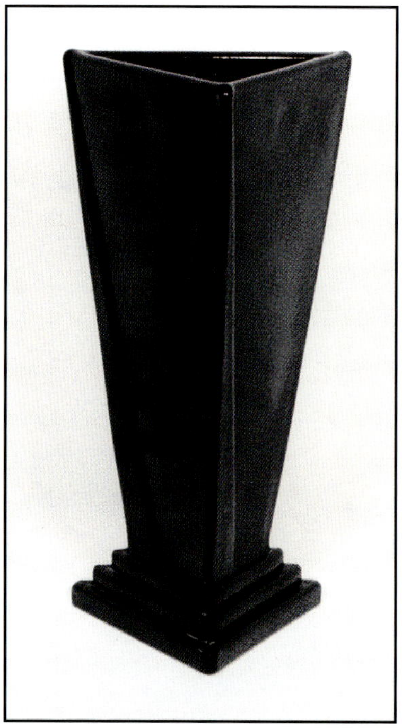

Modernistic pattern triangular vase #33. New Martinsville Glass. Design patented 1928. 8 1/2 inches. (Six 2003) $80-100.

Vase #2425. Fostoria Glass. 8 inches. 1930-1933. (Long & Seate 2003) $80-90.

Art Nude triangular vase. McKee Glass. #100 is 8 inches tall, $140-185; #200 is 7 inches, $120-145. (Stout 1972)

Vases. The form has been attributed by author Dale Murschell to Maryland Glass. Left, undecorated, $14-18; right with gold encrusted band with griffins, $45-65.

Ribbed vases by Fostoria Glass. Left: "Lotus" #2428, 9 1/4 inches, 1930-1940, $60-70; right: #2387, 8 inches 1953-1958, $80-100. (Long & Seate 2003)

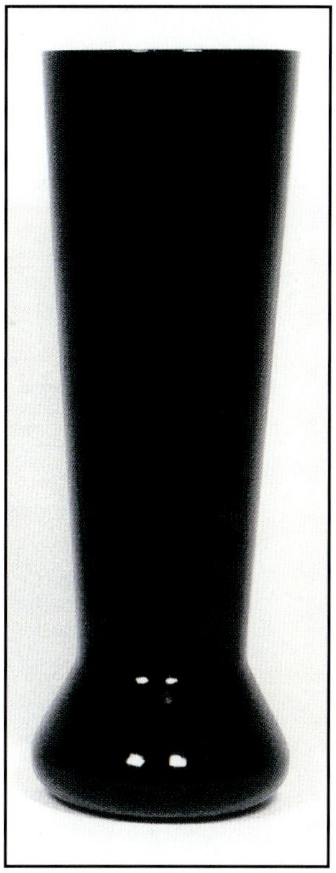

Bulbous base vase. Attributed to Louie Glass. 9 1/4 inches. 1930s. Circa 1930s. $20-32.

Vase from a form originally in the Viking Epic line. 6 1/4 inches tall. Clear vinyl Viking sticker. Circa 1980s. $20-28.

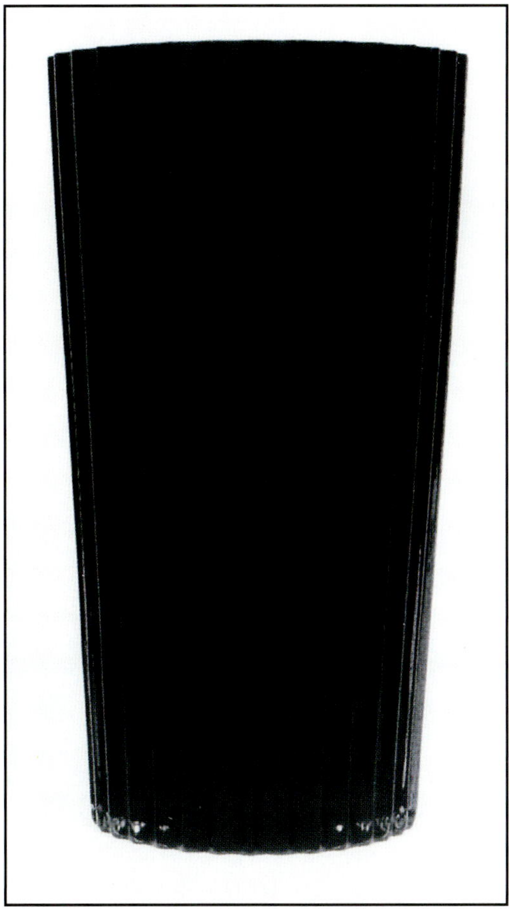

Art Deco Vase #2073. Pilgrim Glass. 12 inches. Cased crystal over black. Pilgrim Glass catalog. $25-35.

Right:
Elaborate Silver Overlay ribbed vase marked Rockwell Silver Co. 7 1/4 inches tall. Part of a complete tableware set. Similar to Fostoria vase forms, but the dimensions are not right for known Fostoria. *Collection of Replacements, Ltd. Museum*. $140-220.

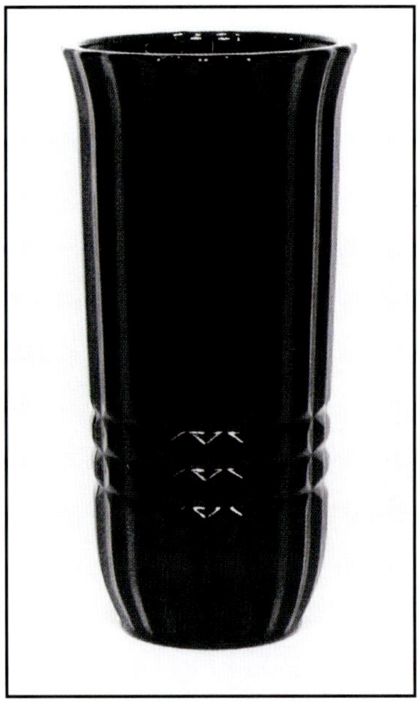

F. T. D. Vase. Contemporary florist ware vase. If it comes free with flowers, it may take a while to gain in value much but many of the objects in this chapter were once inexpensive or free with a purchase. Embossed on bottom "F.T.D.A. © Made in U.S. A.". $4-6.

Vases. Left: New Martinsville Glass, 8 inches, $55-70; right: Fostoria #2409, 8 inches, 1930-1952, $75-85.

Vases. Left: 6 3/4 inches, $30-48; right: 8 inches, $40-60. Westite Glass. Circa 1920-30s. (Six 2003)

Pillow Vase. #56 with Lela Bird etched decoration, #182. Paden City Glass. 8 inches. Circa 1930s. Decoration encrusted with white enamel. (Domitz 2004) $200-250.

Vase. #2404 Fostoria Glass. 1929-1938. 6 inches. (Long & Seate 2003) $60-70.

Elliptical Vase with Lela Bird etched decoration #182, Paden City Glass. 5 1/2 inches. Circa 1930s. (Domitz 2004) $200-220.

Ball Vase. #45 Louie Glass. 5 1/4 inches. Circa 1930s. (Louie Catalog shown in company history chapter.) $10-18.

Ball Vase with collar. Labeled Fostoria Glass. 6 1/4 inches tall. Purchased at Fostoria Outlet in Cambridge, Ohio, in the 1980s. The form appears in no Fostoria catalog. Possibly from a lamp globe mold but left with the bottom not cut out? It is possible it was for a private customer and not part of the Fostoria line. No value established.

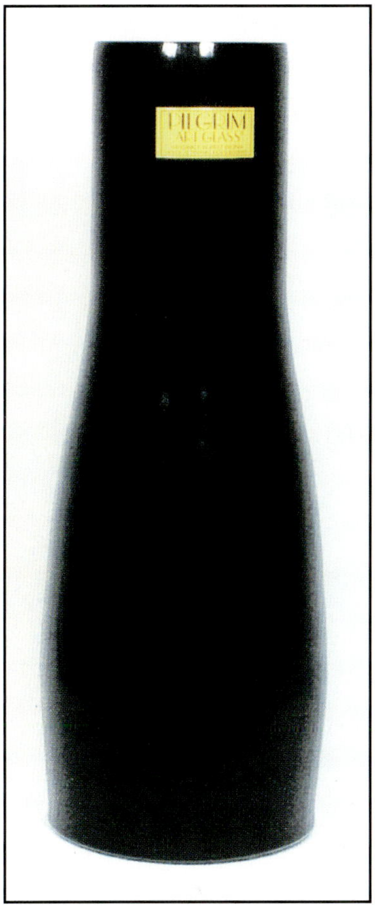

North Pole vase. #2098 Pilgrim Glass. 12 inches. (Pilgrim catalog - WVMAG.) $20-32.

Squatty ball vase. 8 inches. Unattributed. $12-18.

Rib Shouldered Vase. Unattributed. 5 inch oval. Believed to be contemporary. $14-22.

Ball vase. 4 inches. Possibly an undecorated Fenton Art Glass piece from the outlet. This form was produced 1981-1982 as "Silver Poppies on ebony," a silver decorated line. J.C. Penney stores sold the decorated ware. We suggest this may be that undecorated blank. Undecorated as shown, $8-16.

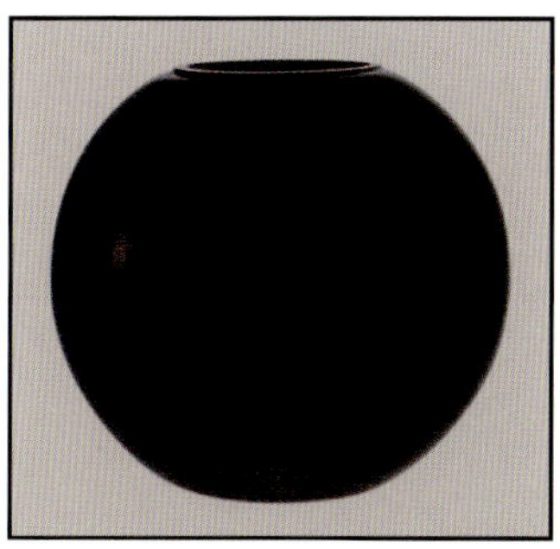

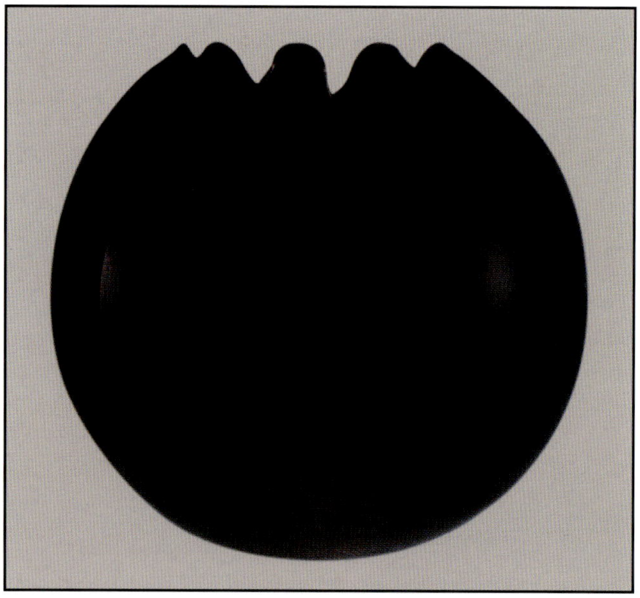

Rose Bowl with crimped top. 7 inches. Unattributed. $16-28.

Crimped Fan Vase #1707. Westmoreland Glass. Oriental Poppy decoration. (Toohey 1999) $34-42.

Fan vase. Black base and ivory bowl. Steuben Glass. Circa 1930s. $950-1,100.

Fan Vase. Central Glass. Paneled under base. 7 3/4 inches. Circa 1920-30s. Same form also shown decorated which is the key to attribution of an otherwise plain and undistinguishable form. (Schmidt 2004) $22-34.

Melon Rib Vase. #857 Fenton Art Glass. 7 1/4 inches tall. Circa 1930s. This mold was used for many forms in many colors over time. Note a similar form with an octagonal sided base was made by L.E. Smith as their #1000 vase. Both companies used the same form to make fan vases, cupped vases, flared bowl-like vases and more. The two are often confused and appear in other books in various colors with misattributions. (Whitmyer 2003) $35-45.

Fan Vase with gold encrusted Dunn's Parrot decoration. Central Glass. 7 3/4 inches. Circa 1920-30s. (Schmidt 2004) $180-200.

Open Work Fan Vase. #310 U.S. Glass. 8 1/2 inches. Circa 1927. Satin. $45-55.

Fan Vase, enamel decoration. Unattributed. The large bulbous knob at the base is the clue to attribution but no matches have yet been found. $18-28.

Double eared Vase. #102/4 with silver band decoration, L.E. Smith Glass. Circa 1920-30s. (Weatherman 1974) $8-14.

Quality sterling silver overlay decoration on vase by L.E. Smith Glass. Hunt scene. Decorator unknown. 8 1/4 inches tall. Circa 1930s. $85-115.

Tut Vase. #2288 Fostoria Glass. 8 1/2 inches. 1925-1940. (Long & Seate 2003) $55-68.

Two handled Vase. #433/4-C L.E. Smith Glass. Circa 1930. 7 inches. Note the same vase is 433/4-C when crimped at the rim and 433/-F when flared out at the rim. (See F.W. Woolworth listing by Smith illustrated in company history chapter.) This is one of the most common of all black glass vases, yet it is deemed as handsome and of interest today as it was in the 1930s. $12-20.

Double handled Vases. L.E. Smith Glass. Left: crimped top #905 undecorated, 10 inches, $28-40; right: smooth top #905 silver band decoration, 10 inches, circa 1920-30s, $24-38. (Weatherman 1974)

Two handled Vase. #15179 U.S. Glass. Circa 1924-1934. 10 inches. $45-60.

Right:
Crimped top Vase with silver decoration. 10 1/2 inches. While no image of this has been found in L.E. Smith Glass literature, it is here placed beside the Smith double handled #905. In light of one of the shown #905 vases having an identical rim treatment and the other an identical silver decoration, it is with confidence we offer the attribution for this vase as a Smith product. $14-22.

Left:
Double handled two ring band Urn with applied, reeded crystal handles. This form, while having identical reeded handles to the other forms shown here, cannot be found in Louie catalogs or in other Louie colors (Cobalt, ruby, crystal, depression pink or green). Thus we hesitantly suggest it may be a Louie product as well, based only on similarity in form and manufacturing traits. 9 1/4 inches. Circa 1930s. $75-90.

Double handled Small Rings Urn with hand applied, reeded crystal handles. Louie Glass. 10 inches. 1930s. $75-90.

Big Rings Vase and Big Rings Urn. Louie Glass. Both 10 inches. 1930s. Left: plain vase form, $40-55; right with double hand applied crystal handles, $55-75.

Double handled #20 Urn with hand applied, reeded crystal handles. Louie Glass. 10 inches. 1930s. (Louie catalog, see company history chapter.) $70-85.

Right:
Double handled Vases. L.E. Smith Glass. Left: #49 6 inch vase with white string decoration, $12-18; right: #433 7 inches vase Dancing Ladies with white string decoration applied over design, $16-22. (Weatherman 1974)

Double handled Vases. Left: #432 with silver floral decoration and crimped top, 8 inches, $14-20. Note this floral decoration often appears on Smith products and may be an initial clue to attribution. Right: #432 two handled vase with smooth flared top, 8 inches, $12-18. These and the dancing ladies vase were manufactured in what today appears to have been immense quantities by L.E. Smith.

Two handled Vase. Design seems influenced by Art Deco themes. Satin. Unattributed. 7 1/4 inches. $65-75.

Metropolis line #2055 vase, Pilgrim Glass. 14 inches shown. Identical form also made in 8 inch size. Late 1980s-1990s. (Pilgrim catalog - see company history chapter.) As shown, $40-65.

Right:
Camelot Vase #2047 Pilgrim Glass. 20 1/2 inches. (Pilgrim catalog - see company history chapter.) $40-70.

Triangle Vases. Pilgrim Glass. Left: 6 inches, $18-26; and right, 15 inches, $80-140. The larger piece is a part of the Pilgrim Masterworks line. (Pilgrim catalog 1990)

Vase #16308 with decoration #2787-KKG. Beaumont Glass. 16 inches. Circa 1980s. $85-110.

Vase with decoration samples. Beaumont Glass. 14 1/2 inches. These came from the Beaumont factory and apparently were used to illustrate how various numbered decal decorations would look on the glass. Note the floral decorations and hand-numbered decoration designation nearby. No value established.

Vase #4631 with decoration #2708-KK. Beaumont Glass. 12 1/2 inches. Circa 1980s. $50-85.

Cornucopia Corinth Vase #3900/575. Cambridge Glass. Circa 1950s. 10 inches. (Cambridge catalog.) $85-125.

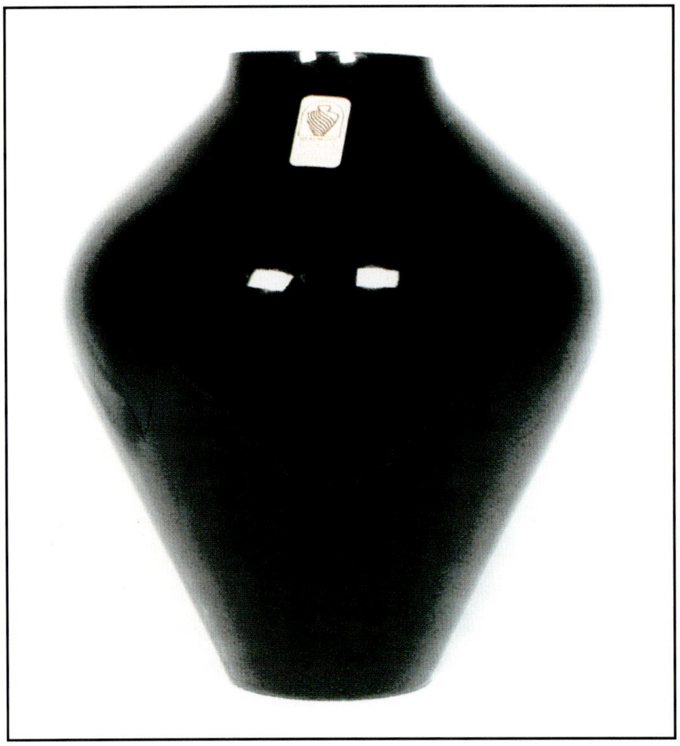

Blown Vase. Beaumont Glass, Circa 1980s. 10 inches. $16-24.

Contemporary import art glass vase. San Pacific Imports. GL 1493 with crystal applied wraps. 10 1/2 inches. Circa 1985. This is one of the better designs and quality products that have flooded the U.S. glass markets in recent decades. It has aided in the speedy demise of the American Glass Industry. Collectors need to be alert to these imports, acquire them for their decorative value, and should not anticipate any increase in value as a collectable for some time to come. (Toohey 1999) $10-16.

Violet Vase, footed. Crystal crimped bowl, black foot. 5 inches tall. This paste mold, hand blown form was commonly produced in several smaller, little-known West Virginia factories. The black foot is less common. It is attributed to Louie Glass based on its provenance, but likely others made identical forms. $8-16.

Contemporary import using black and several colors. Art glass techniques, but very un-American in the final presentation. Again, collecting requires awareness of current worldwide production; and a piece such as this should be purchased for its decorative value, not as an investment or as collectable glass (yet). The use of black glass does not make this "black glass" in the collector's tradition, and we include it here largely to make these points. $8-16.

Trees Vase. Pilgrim Glass. Produced 1987 - 1988 only. Several versions of this line were manufactured. (Pilgrim catalog.) $40-50.

Vase. Barcelona line by Diamond Glass. Circa 1928. Flared rim. $75-85.

Left:
Vase, black ribs and crystal with optic. Fry Glass. Circa 1930. The decoration is commonly called "Ribbon" by collectors. This form is illustrated but not named in the only book on Fry Glass. (Fry Society 1999) $900-1,150.

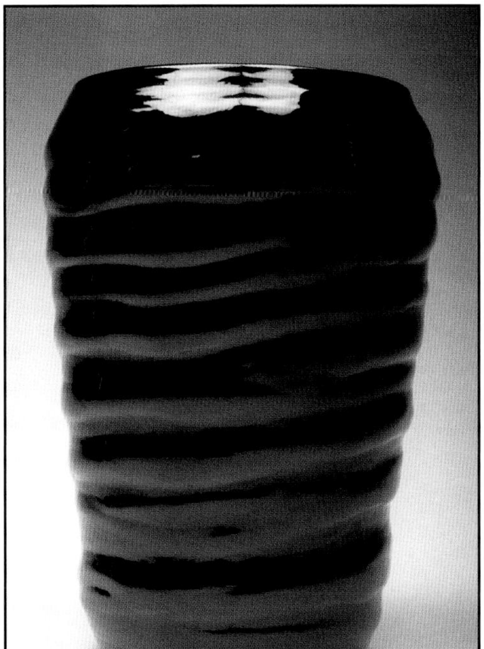

Vase. Barcelona line by Diamond Glass. Advertised in 1928. 7 3/4 inches. (See advertisement in company history chapter.) $75-85.

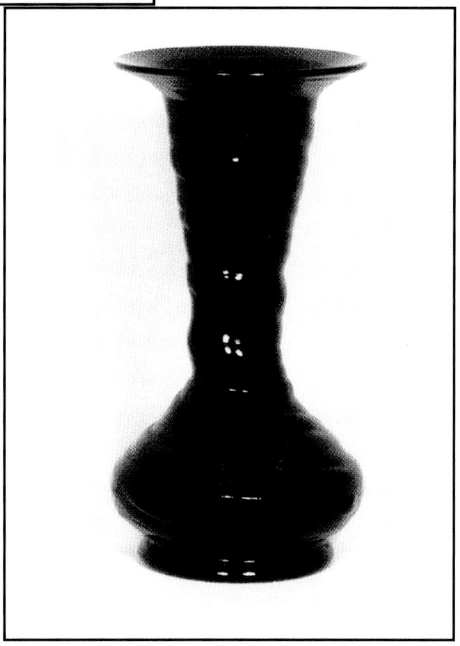

Barcelona line Vase. #5122 Diamond Glass. Circa 1928. 7 1/2 inches. Made to have a hand-wrought, textured appeal. Note: the Barcelona line closely resembles the Catalonia line made by Consolidated Glass at roughly the same time, and the two are very easily confused. We regret that no examples of Barcelona are shown in the book for contrast. (See advertisement in company history chapter.) $40-45.

Barcelona line Vase. Diamond Glass. Circa 1928. Note this is a straight topped version of the vase shown here with the rim flattened. $40-45.

Hobnail Vases. #99 L.E. Smith Glass. Shown 6 inches and 8 inches in height, and with different rim treatments. (Weatherman 1974) Smaller, $8-12 and larger, $8-12.

Vases, detailed pattern in vertical panels. Left: 7 inches, $12-22; right: 5 inches, $10-16. Unattributed. The authors suggested all the usual suspects as possibilities. We even entertained the possibility that they were not American but admit in the end this is unusual and we offer no real conclusions or suggestions.

Empress Vase. #8251 Fenton Art Glass. Note the mold has a history having belonged previously to Verlys of America who produced it only in crystal. Fenton embossed mark on to all Fenton production. 7 inches. Produced in 1969 only. (Walk 2003) $80-95.

Vases, heavily patterned with checker board recessed squares. These two related vases, both 6 inches and with differing treatments to finish the rim, could not be attributed as to manufacturer by us. Either form, $8-16.

Round Vase. Springtime line #1250 Cambridge Glass. 6 inches. Satin and glossy finish. (Cambridge Catalog 1930.) $90-125.

Spooner. Dalzell, Gilmore & Leighton Glass. 4 inches. Single layer of black glass produced in the mold commonly used for the Floradine and Oriental (Onyx) art glass lines. Circa 1889 or later. Uncommon color, perhaps unique. No market value established.

Katydid Pillow vase, 7 inches, $550-650, and Dragon fly vase, 5 3/4 inches, $300-400. Both Consolidated Glass. Circa 1926. (Wilson 1989)

Vase #700 line. Consolidated Glass. 10 inches. Circa 1926. (Wilson 1989) $1,200-1,500.

Love Bird Vase. Consolidated Glass. 10 1/4 inches. Circa 1926. (Wilson 1989) $900-1,400.

Poppy Vase. #16256 U.S. Glass. 6 inches. Circa 1920s. Satin. $38-44.

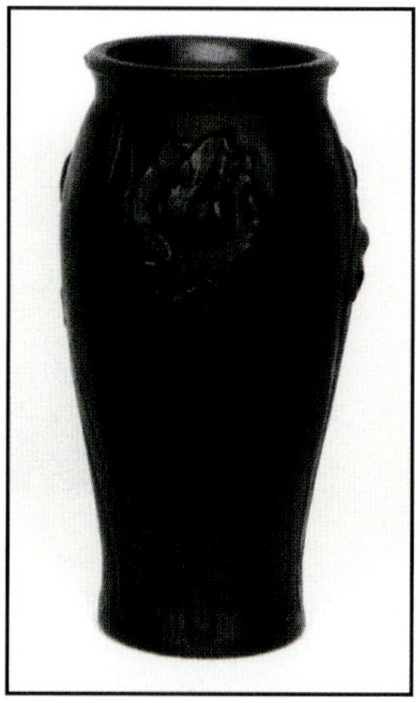

Iris Vase. #16254 U.S. Glass. 6 inches. Circa 1920s. Satin. $55-65.

Pine Cone Vase. Consolidated Glass. 7 inches, satin. Circa 1926. (Wilson 1989) $550-600.

Aster Vase. #16273 U.S. Glass. 6 inches. Circa 1920s. Satin. $40-55.

Poppy Vase #16255. U.S. Glass. 8 5/8 inches tall. Satin with red and green decoration. Circa 1925. As shown, $50-65.

Regina Vase. #210 Paden City Glass. With Black Forest etched decoration. 10 1/2 inches. Circa 1930s. $220-250.

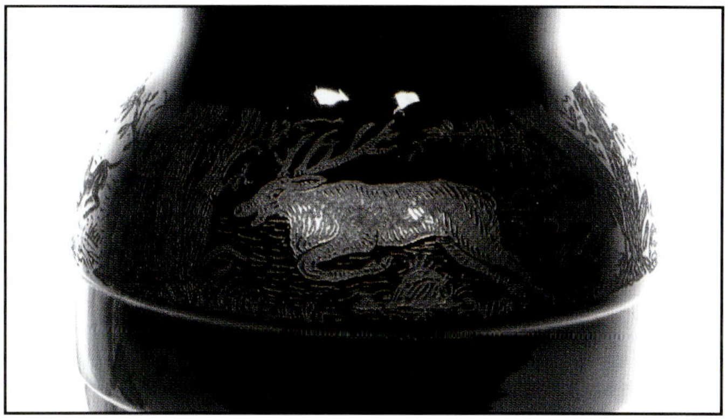

Close up of the Black Forest plate etching on the Paden City vases shown here and line of other pieces.

Poppy Vase. #16255 U.S. Glass. 8 5/8 inches tall. Satin, otherwise undecorated. Circa 1925. As shown, $40-46.

Regina Vase #210. Paden City Glass. With Black Forest etched decoration. 6 1/2 inches. Circa 1930s. $120-145.

Frances pattern crimped large vase #2010. Central Glass. 10 inches tall. (Schmidt 2004) $130-155.

Wall or pocket vase #1681. Fostoria Glass. 1925-1927. (Long & Seate 2003) $75-90.

Vase. #184 Paden City Glass. With etched decoration Lela Bird. White shoe polish to show etching. 10 inches. Circa 1930s. (Domitz 2004) $220-250.

Vase #782. Cambridge Glass. 9 inches. Note original paper label intact. $50-65.

Shouldered vases. These two, very plain but seemingly related, vases, 7 inches and 9 inches, have eluded us, and no attribution is made. Smaller vase, $6-10, larger, $8-12.

Shouldered vase. Hand blown, hand finished top. 6 1/2 inches. Unattributed. $8-12.

Ribbed vase #45. Louie Glass. 11 inches, crimped top. 1930s. Note: this example has a silver paper label declaring it to be "Souvenir of Parkersburg W.Va." (Louie catalog, see company history chapter.) $35-48.

Shoulder vase. Top has ridge, not hand finished. 8 inches. Unattributed. $8-12.

Balustrade vase #23. Louie Glass. 12 inches. Crimped top. 1930s. (Louie catalog see company history chapter.) $30-42.

Crimped top vase. This simple form remains Unattributed. The crimp is common, but the base suggests it was held at the bottom by a crimp (hand tool) during the process of crimping the top. This would be a strong clue as to its place of origin. It is not believed to be in the style of central West Virginia glasshouses. 8 inches. $8-12.

Bulbous vase, "egg in sand" design band near rim. 6 inches. L.E. Smith Glass. Circa 1930. (See Smith Glass illustration in company history chapter.) $10-14.

Bulbous vases, Greek key design near rim. 6 inches. These hand-painted vases will earn an attribution of L.E. Smith Glass here based on the appearance of the identical form appearing in Smith Glass literature but with an "egg in sand" decoration in the embossed pattern band. This is conjectural for us as an attribution for the present time. $10-16 each.

Low, bulbous vase with crimped top. Louie Glass. Circa 1930s. 5 inches tall. $12-20.

Left:
Rose design vase #4/108. Imperial Glass. Crimped top. Circa 1953. (Imperial Glass Co. catalog #53.) $22-34.

Right:
Vases, 6 inches. One is identical to that shown nearby but without decoration. The other offers yet a third decoration in the rim band where the Greek key or "egg in sand" designs also appear. Undecorated, $6-12.

Vase with silver deposit decoration. 8 3/4 inches. Unattributed as to manufacturer of glass or decorator. $80-120.

Ribbed and crimped vase. Louie Glass. Circa 1930s. 7 1/4 inches. $20-28.

"Sac" Vase #1845. Pilgrim Glass Corporation. 6 inches. Note: the "Sac" form, sold with cord on it, made in both 6 and 8 inch sizes, and probably both sizes in black though not verified. (Shape appears in 1990s Pilgrim catalogs.) As shown, $10-18.

Oval vase #1228. Cambridge Glass. 9 inches. Circa 1930s. $75-90.

Crimped Vase #7696. Fenton Art Glass. 7 1/4 inches. 1985-2001. (This appears to be an item in the catalog only with hand painted decoration and thus likely this undecorated item was a Fenton Gift Shop item.) $20-30.

Sophisticated Ladies sand carved vase #7561SX. Fenton Art Glass. 10 3/4 inches tall. Part of a group of three vases produced under an agreement with the Duke Ellington project Partnership. The sand carved decoration was adapted from a popular musical playbill. Also in the line were a round vase and a cylinder vase. 1982 only. (Fenton 1982 catalog.) $200-250.

Left:
Crimped top vase #7690. Fenton Art Glass. (Walk 2002) $18-26.

#809 Marquis vase. Tiara / Indiana Glass. 6 1/2 inches tall. 1989-1993. This exceedingly art deco designed vase is a reissue of Indiana's #303 pattern, originally introduced 1936. (Teal 2000) $18-26.

Vase, left, and salt shaker. No dimensions given. Both attributed to Maryland Glass by author Dale Murschell. Vase, $8-12, salt shaker, $6-10.

Left:
Nudes in relief tri-crimped vase #1931. L.E. Smith Glass. 7 inches. Circa 1930s. (Weatherman 1974) $18-26.

Four footed, corner ribbed vase. 5 1/2 inches. The style and manner of production strongly suggests L.E. Smith Glass, but no documentation was found. $12-18.

Bulbous bottom crimped vase. 5 inches. Attributed to Westite Glass or its predecessors in Weston. Attribution based on other colors in which this form is found being identical to the colors and treatments (satin) used at Westite. We also were influenced by the frequency with which they appear in the town where Westite was located. Circa 1920-early 1930s. $10-18.

Left:
Hourglass shaped vase #1006. Morgantown Glass Décor line. Circa 1970s. 10 1/2 inches. (Snyder 1998) $24-38.

Corset shaped vase #1213/14. Morgantown Glass. Marketed as "Winken and Blinken candlelight base" circa 1968. While it was produced as a candle base, it is commonly viewed as a vase, and it is included in this chapter because it seemed the logical place to look for the form. (Snyder 1998) $30-35 for base as shown.

Flared rim tapering vase. This form was produced by several glass houses: Fenton, Beaumont, Diamond, and others. We offer no attributions for the form at this time. $12-20.

Crimped vase on four legged base. Beaumont Glass. Gold stork decoration. Circa 1930s. Note this base and decoration allow us to make attribution for this vase. The form resembles one shown here without the crimp and representing the most common variety. As shown, $180-200.

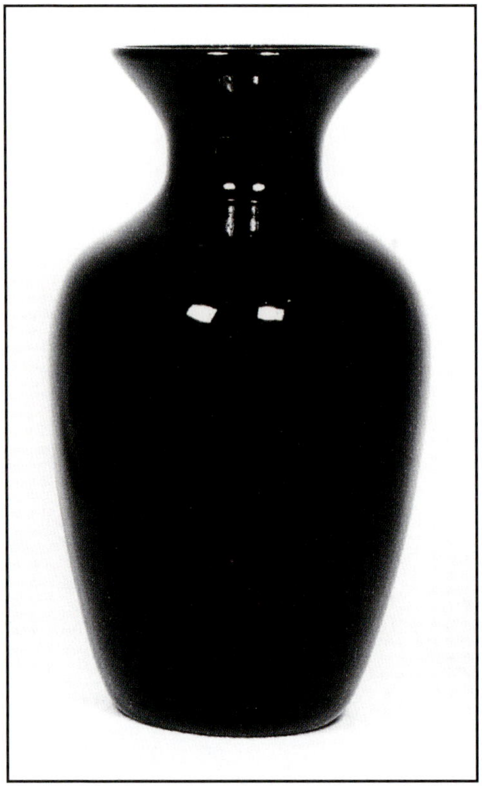

Shouldered vase #91. Louie Glass. Circa 1930s. 8 inches. (See company history chapter for illustration of same vase without flared rim.) $10-14.

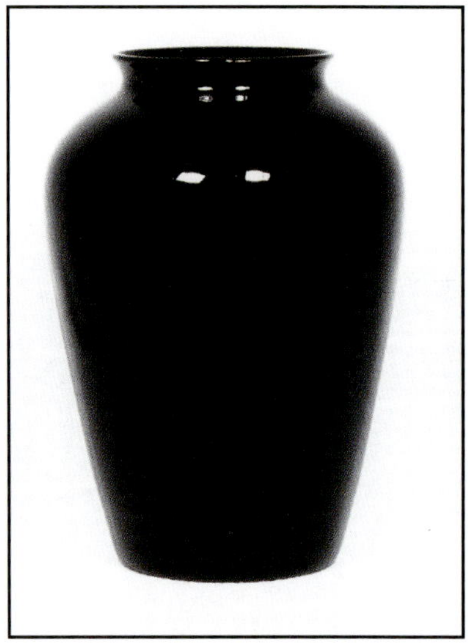

Vase #2. Louie Glass. 6 1/2 inches. Circa 1930s. (See catalog page in company history chapter.) $8-12.

Fine ribbed hourglass shaped with crimped top. Louie Glass. Hand enameled decoration. Circa 1930s. Same form made later in other colors by machine. $18-26.

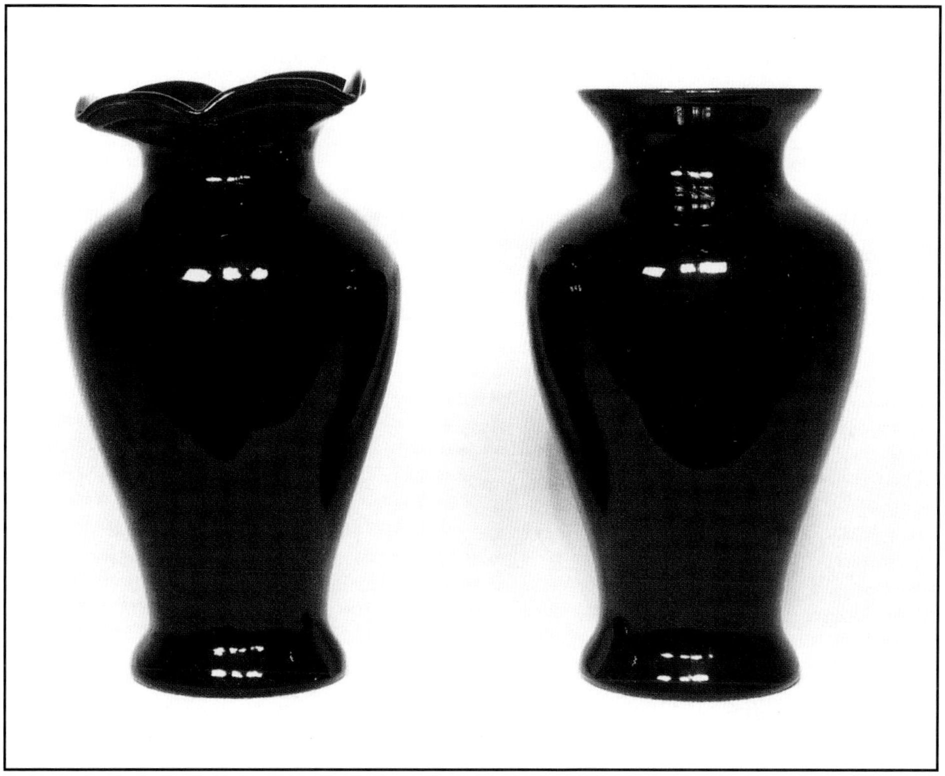

Shouldered vases. Left: crimped top, $12-20; right: flared top, $10-16. Both 9 1/2 inches. Unattributed but hand-blown based on finish to rim. $12-20.

Vases. Left: 7 inches, $8-12; right: 9 inches, $10-16. Not hand-blown due to mold seams appearing on lip. Commonly attributed to L.E. Smith Glass, but no documentation found. Circa 1930.

Vases. Left: 9 inches, $10-14; right: 11 inches, $10-18. Not hand-blown due to mold seams appearing on lip. Commonly attributed to L.E. Smith Glass, but no documentation found. Circa 1930.

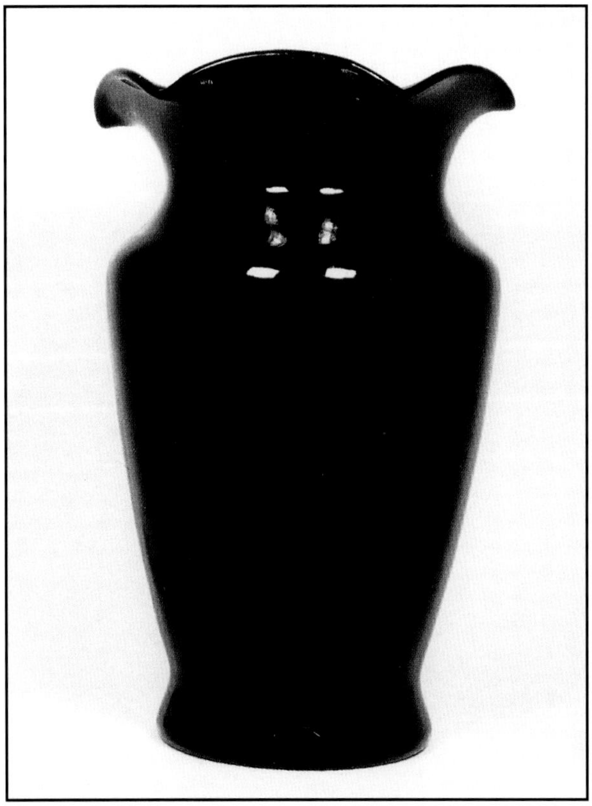

Crimped top vase. #103 McKee Glass. 11 inches. Circa 1920-30s. Commonly found in other opaque colors as well. (Stout 1972) $18-28.

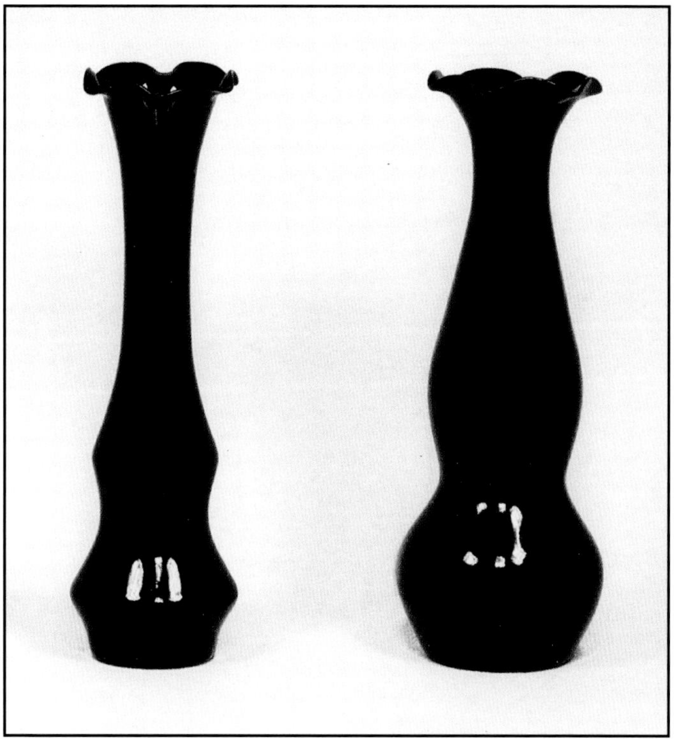

Crimped top vases. Left: #58 Louie Glass. Right: Unattributed. Both crimped top and both 8 inches. (See company history chapter for Louie vase catalog illustration.) Either, $8-14.

Bud vases. This is one of the most common forms of hand-blown bud vases made for decades in perhaps two dozen or more American glasshouses. Attribution is difficult to impossible if undecorated. Note the two variances in crimps and slight shape variances in the two shown. The left decorated vase features a simple floral dot often attributed to the Weston Area/Louie companies. Any similar vases, $8-14.

Two handled vase. #2360 Fostoria Glass. 10 inches. 1930-1940. (Long & Seate 2003) $65-75.

Pressed glass paneled vase. Rayed bottom, flared at rim. Silver decoration: "Rex New Orleans 1914." 6 1/4 inches. Unattributed. $26-36.

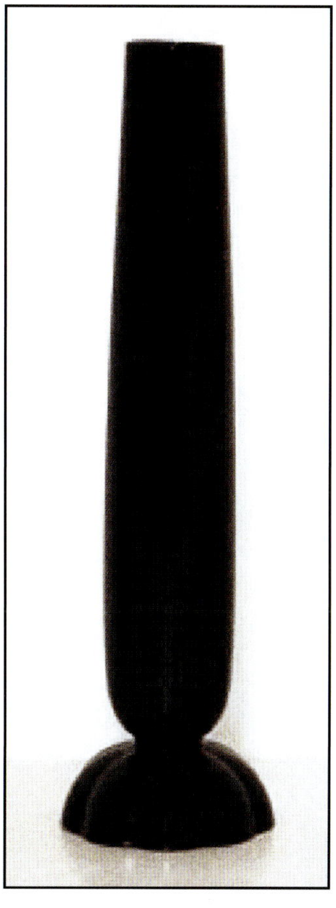

Bud vase #5088. Fostoria Glass. 1931-1939. 8 inches. (Long & Seate 2003) $28-34.

Chapter Thirteen
Novelties, Whimsies, and Miscellany

Slippers. Fenton Art Glass. Left: Hand painted 5 inches signed "hand painted by S. Hopkins" with Fenton logo, \$18-24. Right: hobnail slipper, 6 inches, \$10-18.

Boot with Spur. 3 inches tall. Sowerby or Challinor and Taylor product as we believe both made similar objects. Circa 1900. \$30-44.

Rose Slippers. Fenton Art Glass. Left: 6 inches signed "hand painted by K. Lauderman," \$18-24; right: Rose slipper, undecorated, \$10-18.

Hobnail toothpick #3795 BK. Fenton Art Glass. 3 inches. Mid-1970s. \$10-12.

Swan toothpick #SW-2. Westmoreland Glass. 2 1/2 inches tall. Also marketed as a cigarette holder. $18-28.

George Washington 1732-1932 plate. Plate is one of several souvenir pieces made in 1932 to commemorate the bicentennial of Washington's birth. 8 inches. Attributed to L.E. Smith. $40-55.

Garfield commemorative plate. Lace edge. Note President Garfield was assassinated in 1881 and several objects were made to memorialize his life. This likely is of that very narrow time frame. $75-140.

Forget-me-not hand painted plates. Westmoreland Glass. In the style commonly called Mary Gregory, winter scenes, left No. 2/ 8 inch/ 42 decoration plate and right No. 2/ 8 inch/ 43 decoration. Both appear in 1970s company catalogs. See company history chapter for others in the series. $45-55.

Forget-me-not plate, hand painted No. 2/ 8 inches/ 12 decoration. Westmoreland Glass. In the style commonly called Mary Gregory. Shown here is a girl with dog plate, children were a common theme in this decoration, which uses heavy enamel to simulate more expensive hand carved cameo glass. $45-55.

The Little Brown Church in the Vale Christmas Plate 1970. Fenton Art Glass. The first in an annual series that was issued for several years. Christmas plates and black were not a particularly strong market perhaps. It appears to not have been made in the cataloged line but sold in the gift shop perhaps? 8 inches. With text on reverse. No value established.

Christmas plate. Westmoreland Glass. 1972. Gold decoration shepherds with lambs. Hand painted by Julia Miller. Limited edition of 2,500. $50-65.

Christmas plate. Westmoreland Glass. 1974. Gold decoration Christ kneeling. Hand painted by Mary Anne Smith. Limited edition of 1,500. $55-75.

Christmas plate. Westmoreland Glass. 1973. Gold decoration Nativity. Hand painted by Mary Anne Smith. Limited edition of 3,500. $40-60.

Christmas plate. Westmoreland Glass. 1975. Gold decoration Moses and the ten commandments. Hand painted by Louise Plues. Limited edition of 2,500. $50-65.

Pen holder. 3 1/4 inches. $20-24.

Sengbusch inkwell in the form called "Observatory." 4 1/2 inches. $65-95.

Two part pen holder ink well. Marked "Property of U.S. Navy." 4 1/4 inches wide. Note the pen holding ball rotates or is removable from the base. $45-55.

Schaffers marked desk pen holder. Marked "Schaffer's 505H Cartridge Fill Desk Set." 4 x 4 inches. Shown with replacement pen, $25-30.

Desk tray. Three indents for bookkeeper's ink bottles of red, black, and blue inks plus depression to rest pen. 7 inches long. $18-26.

Figural Bottle Club paperweight. 4 1/4 x 2 3/4 inches. While it proved difficult to photograph the top reads "This certifies that _____ is charter member no. ____ of the Figural Bottle Club 1972-1973-1974 Thomas J. Smith executive director" with the name appearing in script. The blanks were meant to be completed by engraving a member's name once they subscribed to a limited edition, collectable bottle series. The reverse is embossed with the Big Pine Key glass company triangle surrounding "BPK". $20-24.

Figural seals base. 3 1/2 inches tall. Unattributed, this base held a small dime store fish bowl. As shown, $12-20.

"Pres. John Fitzgerald Kennedy 1917-1963" embossed paperweight. Unattributed. Likely circa early to mid-1960s. 3 1/2 inches. $30-34.

Five legged base. 5 inches diameter. Fenton. $65-85.

Horses Cameo Glass paperweight. Pilgrim Glass Corporation. Blown black over "mocha" colored glass. A popular, good selling item in the Pilgrim Cameo line. Bottom engraved "913142 Kelsey/Pilgrim 2000 TK." $250-380.

Decorated white bowl. Beaumont Glass. Fer-Lux is the opaque white color's name from Beaumont and shown here with the proper black four footed base, two pieces. $225-250.

Black glass bases, made to hold and accent bowls of various colors, were popular in the 1920-30s. Each glass company manufactured theirs with a slight, or not so slight, difference from others. Matching the original base to a bowl is critical to obtain top dollar for the two piece sets. Unattributed here and with values ranging from $5-20. WVMAG owns this study collection and they can be viewed at the museum.

Dragon sand carved egg. Pilgrim Glass. 3 inches tall. Pilgrim produced a series of sand carved eggs that appeared in the catalogs of 1994, '95, and '96. Kelsey Murphy, Pilgrim art director and designer, reports that neither black eggs nor the dragon design appeared in the cataloged line. This sample was purchased at the Pilgrim Gift Shop and appears to be an experimental piece. $40-48.

Egg on pedestal #5145. Fenton Art Glass. Signed on bottom "hand painted by L. Emerson No __ out of 2500." 3 3/4 inches tall. $25-35.

Pressed egg left on crack off. Fenton Art Glass. This egg retains the base which is a part of the hand pressing process, but is intended to be "broken" off and ground flat at the bottom of the much smaller decorative base. These eggs in various colors are sold at times in the Fenton factory gift shop in Williamstown, West Virginia. $30-38.

Glass eggs. Pilgrim Glass. Black cut to clear. Left: owl through window with moon and right: boy with clouds. 4 inches tall. Part of a series looking through the black into the crystal silhouetted window. Kelsey Murphy designed, illustrated, and cut each of these herself as the beginning of a series remembering childhood stories and illustrations. They were made in limited quantities. *West Virginia Museum of American Glass collection, gift of the artist.* $350-390 each.

Lamb mug. Attributed to Iowa City Glass Co. Circa 1881. $80-110 in mint condition, not as shown.

Lacy Medallion pattern mug. U.S. Glass. Originally named Jewel by the manufacturer. 3 3/4 inches height. Souvenir of Indianapolis, Ind. and hand decorated. Note: Souvenir decoration may increase value in that area but generally adds little to value. Circa 1890s. Found more often in other colors. $50-65.

Mason's Patent November 30th 1858 with CFJCo monogram fruit jar. Half gallon size. Note that some original black glass canning jars do exist, but they are truly rare. If you see one and it seems too good to be true … well. Almost all black glass fruit jars found will be contemporary reproductions for the collector or decorator market and have marginal value beyond their decorative appeal. $12-22.

Soda cup holder. This base was manufactured for the "VEECUP" line of soda fountain supplies marketed from Chicago, Illinois. It is so embossed on the base. A paper or other cup insert fits into the black glass base. $20-28.

Glass greeting card. "RUMOR HAS IT ... SO HAVE YOU" Produced on a dark gray flat glass created for heat and light resistant windows. This is but one example of an extensive line of glass greeting cards manufactured by Houze Glass of Pt. Marion, PA and marketed in mailable boxes. The image of the card determines value. As shown, $4-8.

Toy mug. 2 1/2 inches in height. $8-12.

Close up of the embossing on the base of the cup holder shown above. Note that a hole goes completely though the base.

Fenton logo placard #8599 E3. Fenton Art Glass. These octagonal based signs feature the contemporary Fenton logo, and were made for dealers as display pieces and as special limited productions for Fenton collector clubs. Produced 2001. 5 inches in width. $25-35.

Miniature Gazing Globe for Zeller Glass. This hollow ball is mirrored on the inside and fitted into a black glass base. They relate to the larger gazing globes used in gardens. The bottom of the base may read "MFGD BY CHAS. F. ZELLER MORGANTOWN W.VA. USA" or it may be unmarked. Zeller had the components made for him (private mold work in the glass industry) and employed young ladies to silver the interior of the ball and affix it to the base. 4 3/4 inches tall. Circa 1940-1960s. Widely marketed to large national chain dime and variety stores. $20-30.

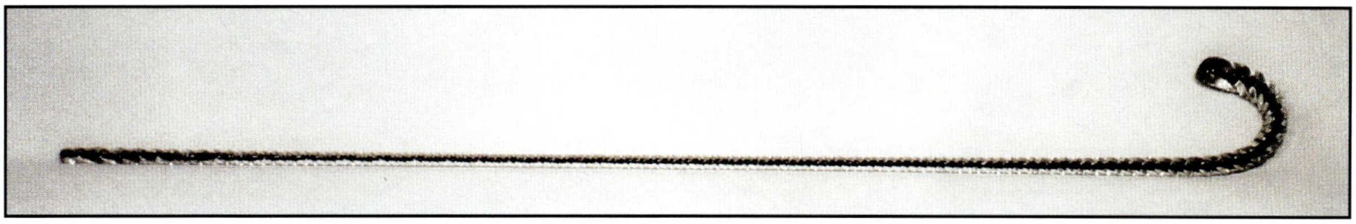

Black filament encased in crystal glass cane. Glass canes were showpieces of the glass artisan's skill and were historically carried by marching glass workers in Labor Day parades. This beautifully spiraled cane is one of many believed to be unique to the hand crafting of John Pertz. The Pertz family was involved in the ownership of West Virginia Glass Specialty in Weston, WV, where the canes were crafted. The encased filament can also be found in cobalt or ruby. 43 inches. $200-280.

Waste bin? Unknown novelty. Unattributed. 3 1/2 inches height. $12-18.

Canoe novelty. Unattributed. U.S. Glass version is shown in a 1926 catalog (Tiffin Club 2004), has a smooth rim, and this variation has a rope edge rim causing us to take caution in saying U.S. Glass. 6 inches length. $20-32.

Spittoon. 4 3/4 inches tall. Glass Spittoons are relatively uncommon forms. Unattributed. $60-85.

Pistol. Westmoreland Glass. Hollow, open at muzzle. 5 inches in length. A glass pistol was also made by Cambridge and later at Beaumont Glass. $75-85.

Miniature wash boiler novelty. Westmoreland Glass. Marked on base "WG". 4 inches length. $18-28.

Stand for Keg decanter. Cambridge Glass. This rustic log patterned stand was made to hold a oak keg styled decanter. 5 inches in length. Circa 1930s. $20-35.

Candleholder (left), $12-20 and vase or sugar bowl, $10-16 in hand blown glass with crystal handles. The bottom of these pieces bear a rough pontil scar. These forms reflect those popular at a number of smaller handmade glass houses in the U.S. at mid-twentieth century and scholarship has yet to document or sort them. It remains a challenge for a future generation.

Ring holder. U.S. Glass. Called "Milady Ring Holder" in an advertisement appearing in *China, Glass and Lamps*, May 30, 1922. (Tiffin Club 2004) $22-38.

Chapter Fourteen
Company Histories with Illustrations

Akro Agate Company

Clarksburg, West Virginia
1914-1951

This firm began in Akron, Ohio, (thus the "Akro" in the name) in 1911 before relocating to Clarksburg, West Virginia, in 1914. The dominant product line for Akro was marbles, the small glass spheres for children's games. Significant lines of children's glass dishes were made by the firm as well and, after Westite Glass (see that entry below) burned in 1936, the molds for glass flower pots and planters were transferred to Akro Agate. The "garden" line proved a profitable line at Akro and it will be these forms: flower pots, planters as well as a few compotes, ash trays, small novelties, and covered powder dishes that one might find in Akro's black glass.

Akro Agate's assets were sold to the Clarksburg Glass Company in 1951.

Atterbury & Co

Pittsburgh, Pennsylvania
1859-1899

Atterbury & Co. was founded by Thomas Bakewell Atterbury and located in the Southside of Pittsburgh, Pennsylvania. It became known as the White House Works.

It was well known for pressed glass. Elaborate and skilled mold makers created a line of covered animal dishes, often made in white or milk glass, that have become sought after today. Atterbury also made tableware and kerosene lamps. Many of its products were made in milk white opaque glass, but some items were made in black glass, including a selection of various sizes of children's and novelty mugs.

The Beaumont Company

Morgantown, West Virginia
1916-1991

The Beaumont Company was incorporated in Morgantown, West Virginia, in October of 1916 and ceased manufacturing in September of 1991. Its founder, Percy J. Beaumont, was not new to the glassmaking world. His sister was married to Harry Northwood, a noted glassmaker. Percy Beaumont was responsible for organizing two other glass factories before he settled in Morgantown. The earlier factories were located in Martin's Ferry, Ohio, and Grafton, West Virginia.

Little has been written about the Morgantown factory until recently. The Beaumont Company specialized in the production of lighting ware, lamp parts, and stationery products such as inkwells, pen trays, and sponge cups. Most of this merchandise was sold under the name of the lamp or inkwell company that purchased the glass parts from Beaumont. Wholesaling their glass to secondary manufacturers unfortunately resulted in the Beaumont name not being published or widely recognized. Beaumont did make a line of tableware and decorative pieces in the 1930s. These were marketed at trade shows and little catalog or archival documentation has been found to date. The company also merchandised, under the Beaumont name, a line of art glass pieces, many of cased glass, in the 1980s.

Beaumont's black glass lines touched on most of their areas of manufacturing. Black glass was produced in one line of inkwells made for the Sengbusch Company. It was also used in the larger vases and decorated wares in the 1960s and 1980s. Black glass was used in their production in the 1930s to create their hexagonal shaped luncheon sets, bowls, vases, and Beaumont's unique four-footed base.

Beaumont Company original catalog page circa 1970.

Beaumont Company original catalog page circa 1970.

Beaumont Company images from original company catalog circa 1980s.

Beaumont Company images from original company catalog circa 1980s.

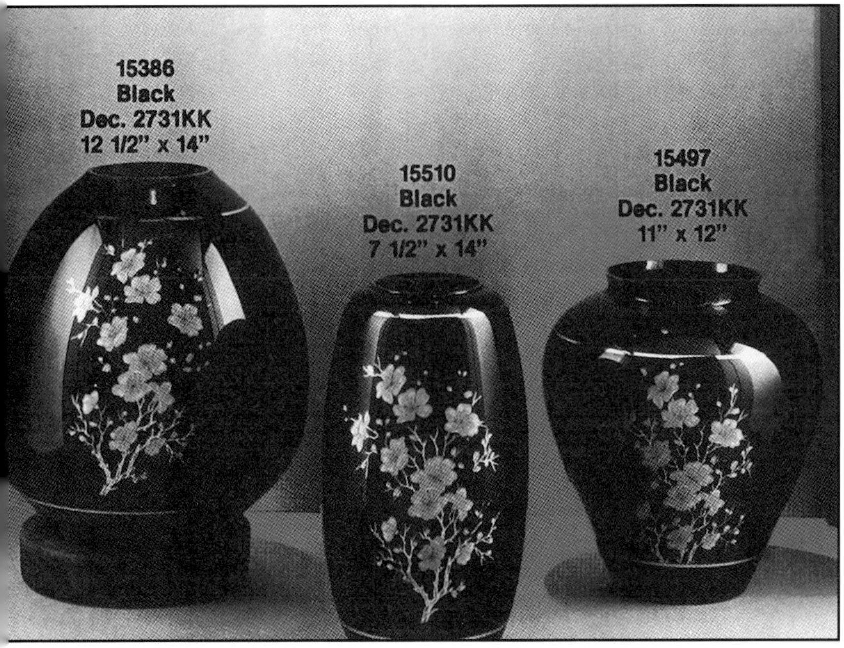

Beaumont Company images from original company catalog circa 1980s.

Beaumont Company images from original company catalog circa 1980s.

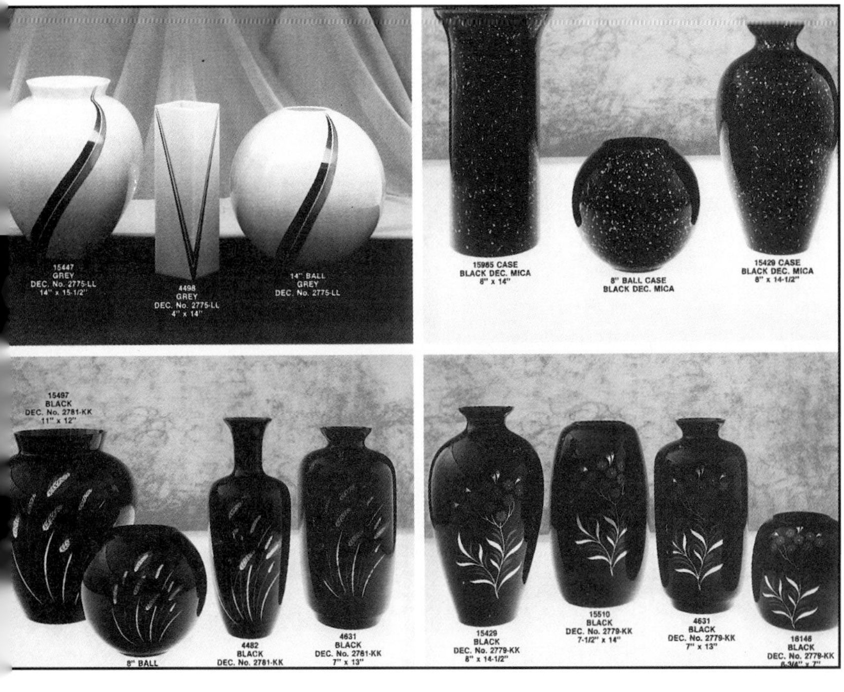

Beaumont Company images from original company catalog circa 1980s.

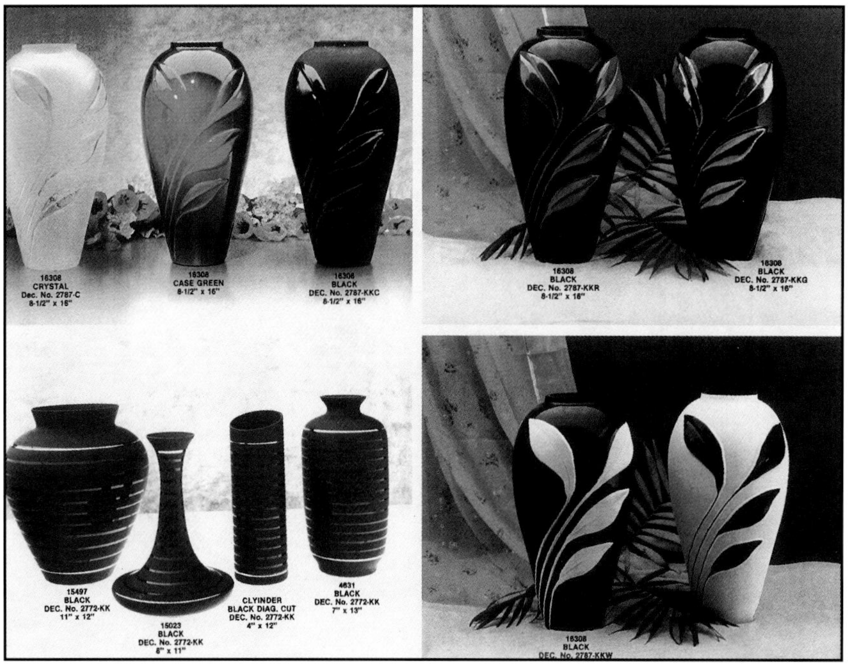

Beaumont Company images from original company catalog circa 1980s.

Blenko Glass Company

Milton, West Virginia
1930 - operating at the time of this writing

English glassmaker William John Blenko founded the company in Milton, West Virginia, after decades of setbacks and failed efforts to establish a glass factory in a variety of other locations. The Milton factory opened in 1921 as Eureka Art Glass. Blenko's son, William H. Blenko, Sr., joined his father in 1923, and in 1930 the company name changed to Blenko Glass Co.

Founded to blow cylinder glass that was flattened in the traditional method, creating colored flat glass for windows and architectural uses, the firm entered the tableware market in the Great Depression to supplement declining sales of glass for building use. Being a blown glass shop, Blenko's products bear the characteristic elements of glass blown by mouth such as a pontil scar on the base from the manufacturing process and telltale seeds or bubbles in the glass. Blenko's designs were modern, designer driven, hand blown, and at times ranging from large to monumental in scale. Floor vases can attain in excess of 40 inches in height or an object can be as unexpected as a life-sized cowboy or formal top hat made of glass.

Black was not a common color in the Blenko scheme of things, appearing only recently. Black glass did not appear as a Blenko offered color until 1999.

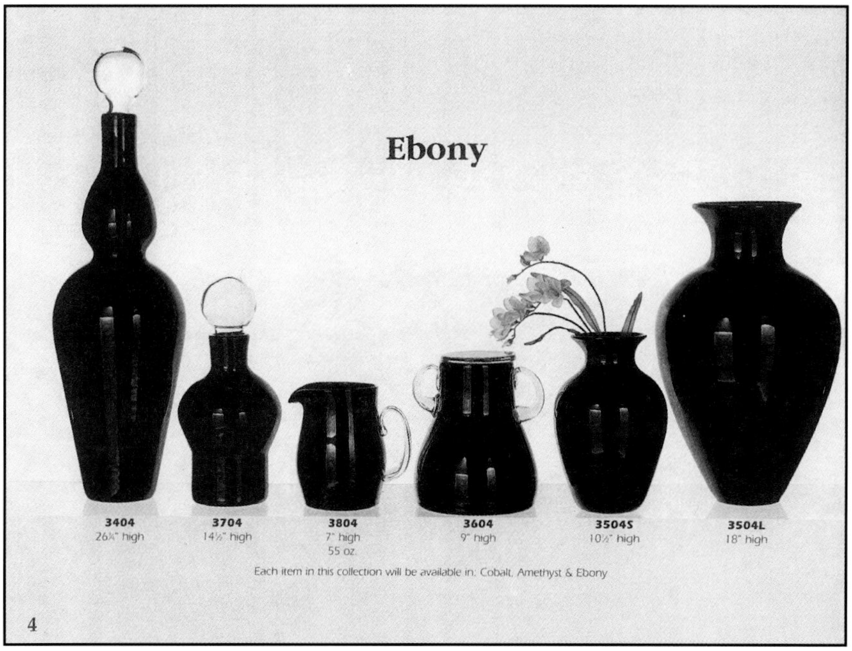

Blenko Glass catalog 2004-2005.

Bryce Brothers Company

Mt. Pleasant, Pennsylvania
1882-1965

The Bryce family had a long and complex involvement in a number of glass companies beginning with James Bryce in the 1840s. The Bryce companies go through many changes in leadership, organization and a number of names over time becoming Bryce Brothers in 1882. The Bryce Bros. Company built a new factory in Mt. Pleasant, Pennsylvania, in 1895. The firm's twentieth century production was largely hand blown lead stemware. The factory remained under family leadership until the 1950s.

Bryce offered both commercial and household glass, including wares used on U.S. Embassy tables around the globe. By the late 1920, colored glass was introduced at Bryce. Production of black seems to have been largely stems and feet on colored or colorless bowls. It seems black continued in the line from the 1920s through the 1950s sporadically.

Lenox acquired a controlling interest circa 1965 and renamed the firm Bryce Crystal for a very short time before it became Lenox Crystal, Inc. in 1966.

Mode stemware line features colored bases in Morocco, Gold, Black, Greenbriar and Crystal. Retail price is $2 ea. Bryce Bros., Mount Pleasant, Pa.

Bryce Bros. Crystal illustration appearing in the trade journal, *Crockery and Glass Journal*, February 1954.

Buckeye Glass Company

Martin's Ferry, Ohio
1878-1896

Buckeye Glass opened in the then abandoned factory of Excelsior (Glass)Works of Sweeney, McCluney & Co in Martin's Ferry, Ohio, in 1878. Harry Northwood (see Northwood entry) became general manager in 1888 for a short time. By the 1890s, labor strife led to erratic production before a fire destroyed the factory in February of 1896.

> "Congo" and "Roseblush" are the names of the two new lines of the Buckeye Glass Co. brought out for the Spring trade, and they are beauties of the first water. The number of the "Congo" is 547 and is made in a jet black, with a very elegant enameled decoration by gold print. The contrast is very rich and effective. The "Congo" is made in a full line of tableware, novelties, and specialties, and is the first and only one on the market and will be a fast seller. From the elaborativeness of the design it would appear that the price will be high, but in reality it is not a high priced one. Gold print is an entire new novelty and has never been done on glass in this or any other country. The "Congo" must be seen to be appreciated. (*Pottery and Glassware Reporter,* January 1, 1891)

A report appearing in the *Pottery and Glassware Reporter* of January 15, 1891, reads as follows: "An original lot of glassware and several lines that depart widely from the routine and beaten track is shown by Mr. Gottschalk of the Buckeye. The man who grows pessimistic and loses hope should not fail to see the display of this firm, as it is full of evidences of progress and shows the versatility, originality and enterprise of the American, when really put to the test. In the 'Roseblush' this firm has not only original blown shapes but new tints and colors, which are at once novel and attractive ... the 'Congo' is a new line of black glass, on which the dusted gold decorations show up to advantage. It is finely designed, of striking and attractive appearance, and really a new departure in the glass industry. The gold enamel on both the above lines is something never before successfully applied to glassware in this country, and we have in them something which in design and workmanship, solidity of decoration, color and shape will compare favorably with the finest imported blown and decorated goods. The salts and peppers in the 'Congo' are a neat acorn pattern, and there is an astonishing variety of forms and colors in this exhibit, which at once delights and inspires."

Such are good examples of the manner and style of late nineteenth century prose in its efforts to praise the black glass and other products of Buckeye Glass.

Cambridge Glass Co.

Cambridge, Ohio
1902-1958

The Cambridge factory was originally built by the National Glass Co. combine, but National never operated the plant.

Cambridge Glass first produced pressed glass tableware and barware, and often these were marked Nearcut. Later production turned to decorative tableware, stemware, vases, and a myriad of other items. Cambridge used a mark, the letter "C" in a triangle, to mark their glass in the period of the 1920s.

In 1922 Cambridge introduced its Ebony (black) glass. *China, Glass & Lamps* wrote on January 15, 1923, that the Cambridge Glass Co., "introduced the 'Ebony' glass last year." Many console bowls were made with black bases. Additional Cambridge black glass items include candlesticks, smoker's items, figurines, and more. In later years, Cambridge had a line of satin glass that has a more coarse finish than most satin glass. This ware feels rough to the touch.

"The 'Carrara' is a brilliant white and stands in strong contrast to the deep black 'Ebony.' A large number of new pieces have been added to the 'ebony' line, making this line one of the most extensive ever offered to the trade." (*China, Glass & Lamps,* January 15, 1923)

The Pottery, Glass & Brass Salesman of October 17, 1929 ran a feature which was headed, "Cambridge Glass Co. Black and Crystal Combinations are Particularly Favored." The article reports as follows: "In the Cambridge Glass Company's Chicago display rooms in the Shops Building, 17 North Wabash Avenue, are shown some of the smartest and newest examples of ebony-colored glassware from the company's factory at Cambridge, Ohio. This lustrous ebony, advantageously placed, easily stands out among the other lines of clear, attractive and colorful glassware."

An illusive black glass item is noted in the following account, "Friends of Cambridge Glass Company have received a Christmas gift from that concern, an attractive little ash receiver of unique design. The receiver is made of black glass and is squat in form. It is in two parts, the lower portion being a well some 4 inches in diameter at the widest part and 2 1/2 inches in height. Its widest portion is near the bottom. Setting into this well is an open cover with four rests for cigarette or cigars, the construction making for easy cleaning and assuring that the ashes will not be scattered around. On the side is a holly trimmed legend in red and gold reading, 'Compliments of the Cambridge Glass Company, Cambridge, Ohio, U.S.A.'" (*Crockery and Glass Journal,* February 1930)

Cambridge Glass original catalog. Circa 1925.

Cambridge Glass original catalog. Circa 1925.

"There are several different treatments in the stemware, but the black and crystal is probably the smartest. The engraved bowl with the black stem shows up strikingly when resting on an ebony plate, and is especially effective when used for contrast on an all-white damask cloth. Some of the odd items shown have a gold etching on the ebony. There are generous varieties of these in such items as ashtrays, cigarette boxes, and cake and sandwich plates. Among the most attractive pieces, however, is the all etched ebony cocktail set consisting of a handsome shaker with a patented silver-plated combination top, instead of the usual screw top, and six etched glass with crystal bowl and etched key border black foot." (*1930s trade journal quote*)

The factory closed in 1958 and the assets, including the molds, were sold to the Imperial Glass Corporation of Bellaire, Ohio.

Cambridge Glass original catalog. Circa 1934.

Cambridge Glass original catalog. Circa 1931.

Cambridge Glass original catalog. Circa 1931.

Cambridge Glass original catalog. Circa 1931.

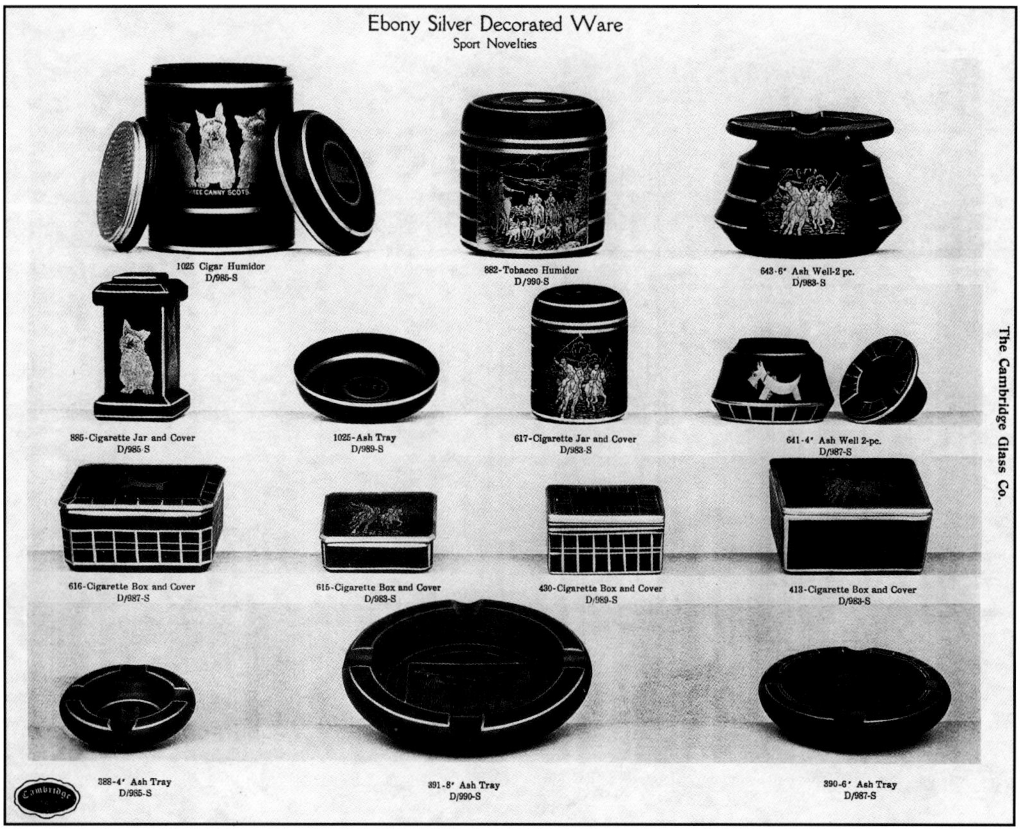

Cambridge Glass original catalog. Circa 1931.

Cambridge Glass original catalog. Circa 1950.

Cambridge Glass original catalog. Circa 1949-50.

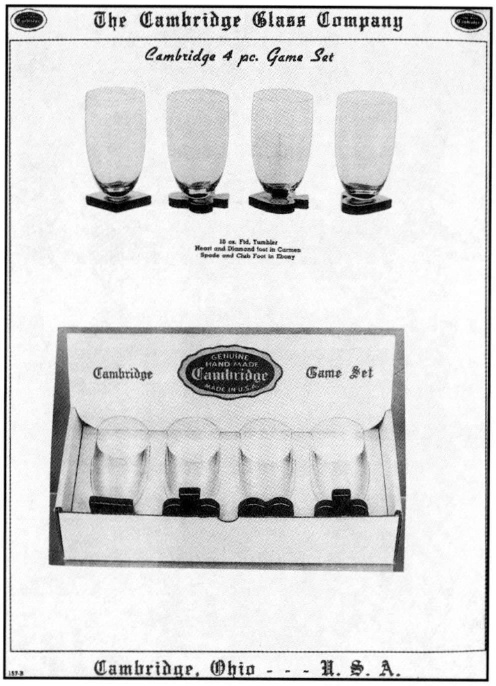

Cambridge Glass original catalog. Circa 1949-50.

Cambridge Glass original catalog. Circa 1950.

Cambridge Glass original catalog. Circa 1949-50.

Cambridge Glass original catalog. Circa 1950.

Canton Glass Company

Marion, Indiana
1903-1958

Canton Glass Company was located in Marion, Indiana, from 1903 to 1958. It relocated to Hartford City, Indiana, and operated there until closing in the 1990s. Canton made both pressed and blown ware, including tableware, accessories, and more commercial objects.

Black glass was produced at Canton in the 1950s. Often black appears in combination with amber in pieces from the Casual line produced by Canton.

Central Glass Works

Wheeling, West Virginia
1895-1939

Central Glass Works was created from the remains of the earlier Central Glass Company after it had been acquired by the conglomerate U.S. Glass. Central Glass Works is the name given the company created when it was sold back to many of the original owners by U.S. Glass. Operating from 1895 until closing in 1939, Central glass of this era included the then popular transparent glass colors, crystal, and black.

As early as 1927 some Central lines appear to have been produced in black as well as other colors. Line 2019, a line of shapes with two open handles, was produced in all black well into the mid-1930s. Some black and crystal combination stemware lines, according to *Central Glass* author Tim Schmidt, continued in production until near the end in 1939.

Central Glass advertisement appearing in *China, Glass and Lamps*, August 6, 1925.

Central Glass stemware offered with green "dishes" from unknown manufacturer. Montgomery Wards Fall & Winter 1930-1931.

Consolidated Lamp & Glass Co.

Coraopolis, Pennsylvania
1894-1964

Consolidated Lamp & Glass Co. purchased the Fostoria Lamp & Shade Co. of Fostoria, Ohio, in 1894. Due to an insufficient supply of gas to operate the Fostoria site, they constructed a new factory in Coraopolis, Pennsylvania, in 1895. For the first five years Nicholas Kopp, an industry figure well known for his innovations in colored glass, managed the factory.

Consolidated Lamp & Glass produced tableware, Martele being one of the most popular patterns, but was best known for its extensive and often elaborate line of kerosene lamps.

Production of black satin glass was begun at Consolidated in 1925.

Co-Operative Flint Glass Company

Beaver Falls, Pennsylvania
1879-1934

The Co-Operative Flint Glass Company of Beaver Falls, Pennsylvania, was founded in 1879 by striking workers from the McKee Glass Company. It was a true co-operative when it was founded, the first glass co-operative in the country. As it prospered, the management changed but it remained an independent company throughout its history, never becoming part of one of the massive combines popular at the turn of the twentieth century. Co-Op, as it is often called, produced a myriad of glass, tableware being its primary product.

China, Glass & Lamps reported on January 15, 1923, regarding Co-Operative Flint Glass Co., that … "Their attraction is in the color treatment which comprises canary, blue, amethyst, turquoise, 'Midnight Black', and green. The 'Midnight Black' also is offered in a number of decorations in gold and silver."

In 1930, the industry press reported "The Co-operative Flint Glass Co., Beaver Falls, Pa., has added a line of Ivy balls in assorted colors to its list of specialties, these balls being in plain and crackled glass. These balls are being placed on the market with black glass bases and with chains." (*Crockery & Glass Journal*, May 1930)

Black glass was advertised in the early 1920s. As there was no reference found to it before that time, it is suggested black glass was introduced around 1920 or shortly thereafter. The Co-Operative Flint Glass Company went into receivership in 1934, a victim of the Great Depression.

Co-operative Flint Glass advertisement appearing in trade journal, late 1920s.

Co-operative Flint Glass advertisement appearing in trade journal, 1928.

Co-operative Flint Glass advertisement appearing in *Crockery and Glass Journal*, December 1928.

Co-operative Flint Glass advertisement appearing in trade journal, 1929.

Co-operative Flint Glass advertisement appearing in *Crockery and Glass Journal*, July 24, 1924.

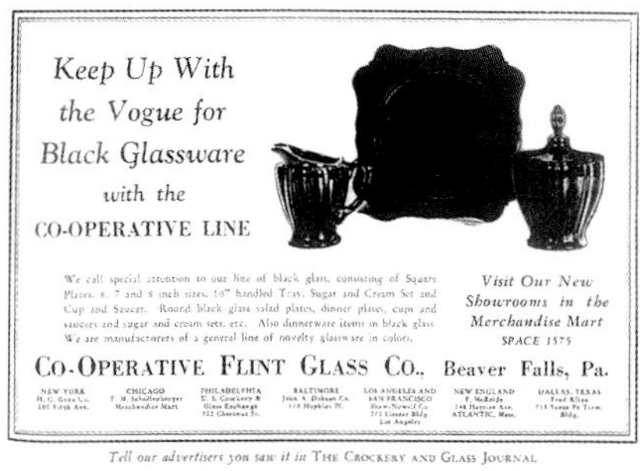

Co-operative Flint Glass advertisement appearing in *Crockery and Glass Journal*, May 1930.

Co-operative Flint Glass advertisement appearing in trade journal, 1924.

Crescent Glass Company

Wellsburg, West Virginia
1908-1983

Three generations of the Rithner family worked here and managed this Wellsburg, West Virginia, glass company that opened in 1908. Crescent made lamps, lamp parts, tumblers, and barware, novelties, candlesticks, and a variety of other forms in an amazing variety of glass colors and treatments.

In 1983 the name was changed to Brooke Glass. The firm operated under that name until 2001 when hot glass production was discontinued; but work as a glass decorating shop continues at the time of this writing.

Crystal Glass Co. advertisement appearing in *Pottery Gazette and Glass Trade Review*, May 1957.

The Crystal Glass Co.

Knottingley, Yorkshire, United Kingdom
circa 1950s

The authors admit little knowledge about The Crystal Glass Co. of Knottingley, Yorkshire, UK. Discovered in literature searches were two illustrated ads (dated 1957 and 1959) from an English tableware trade journal. These illustrate black glass produced by this firm. The advertisements note the proprietors of the company were "Bagley & Co. Ltd." The representatives who distributed this line were from England, Scotland, Australia, Ireland, and Canada. This insures some of the product can be found across the globe and surely on the U.S. markets, having come from Canada and beyond.

Crystal Glass Co. advertisement appearing in *Pottery Gazette and Glass Trade Review*, May 1957.

Dalzell, Gilmore & Leighton

Findlay, Ohio
1888-1899

Beginning in 1883 in Wellsburg, West Virginia, this company relocated in 1888 to Findlay, Ohio, to participate in the great natural gas boom promoted in northwest Ohio in the late 1880s.

Dalzell, Gilmore & Leighton produced tableware, stemware, and kerosene lamps. George Leighton's creation of Oriental glass (now known as Findlay Onyx) is one of the firm's most notable patterns and products.

Black glass production at Dalzell, Gilmore & Leighton was limited, but some lamps with black bases are known, especially with Oriental (onyx) fonts. Other black glass items include the black satin heads found on bottles modeled after a black man and woman- called Snowball and Mrs. Snowball in original ads. A few very rare pieces of Oriental (onyx) are known in black.

The company joined National Glass, a combine, in 1899. Production continued for a few years, but the company was dismantled and molds for some patterns were sent to Indiana Tumbler and Goblet Co. in Greentown, Indiana. Lamp molds were sent to Ohio Flint Glass in Lancaster, Ohio. Molds for new patterns being developed at Dalzell, Gilmore & Leighton were moved to the new Cambridge Glass Co., in Cambridge, Ohio.

Dalzell, Gilmore & Leighton advertisement appearing in *Crockery and Glass Journal*, October 23, 1890, and appearing weekly for several months.

Dalzell, Gilmore & Leighton advertisement appearing in *Crockery and Glass Journal*, December 4, 1890, and running for many issues thereafter.

Dalzell-Viking

New Martinsville, West Virginia
1987-1998

Viking Glass closed its doors in 1986. Shortly thereafter Kenneth Dalzell bought the factory and opened it as Dalzell-Viking. Kenneth Dalzell was then most recently affiliated with Fostoria Glass. He is the great grandson of W.A.B. Dalzell, one of the founders a century before of Dalzell, Gilmore & Leighton (see that entry) and thus represents one of the families whose involvement with glass is the longest in the U.S.

Dalzell operated Dalzell-Viking using the staff, patterns, and other assets of Viking, which had in many ways continued to use New Martinsville Glass assets and patterns a half century earlier. The result is similar or identical to objects made by two and three companies spanning almost a century. Many forms from New Martinsville Glass, including the animal molds, were produced under the Dalzell-Viking leadership in black glass. Some objects bear acid-stamped marks saying "Dalzell." Many do not. The use of the same mold might make distinguishing crystal animals from the 1940s and 1990s difficult. Black glass animals from either New Martinsville Glass or Viking would be uncommon to rare depending on which animal. Black glass animals at times filled many of the shelves in the Dalzell-Viking outlet in the 1990s. Black glass was commonly used at Dalzell-Viking to make votive cups, tableware, animals, candlesticks, and much more. Patterns whose molds had come to Dalzell-Viking, such as Candlewick pattern molds from Imperial Glass, were issued by Dalzell-Viking in black and other colors which would be most unusual for the original manufacturer. These will surely cause collectors difficulty in the years and decades ahead.

In 1998 Dalzell-Viking closed its door and ceased production.

Diamond Glass-Ware Company

Indiana, Pennsylvania
1913 - circa 1931

Diamond Glass-Ware Company opened in 1913 in the factory built in 1892 by the Indiana Glass Company. After the failure of Indiana in 1894 the plant was leased to Harry Northwood. Northwood operated it as the Northwood Company until its acquisition by the National Glass Company in 1899. In 1904 the National sold the plant to the Dugan Glass Company. The Dugan brothers, Thomas and Alfred, were among the first to produce iridescent glassware (known today as carnival glass). When the Dugans resigned in 1913 the Diamond Glass-Ware Company was formed.

Diamond continued to specialize in iridized finishes on glass, as well as light cuttings and other decorative techniques. In 1922, they augmented the line with the addition of the Egyptian Lustre, which featured brilliant iridescent colors on black glass. *The Crockery and Glass Journal* described Egyptian Lustre in their July 20, 1922 issue, stating that it "could also be appropriately termed peacock ware, a very rich and beautiful iridescence ... achieved in combining the shimmering shades of bronze, blue, green, and ruby. Among the items of this fancy assortment are deep bowls, cupped flower bowls, candlesticks in a varied assortment of sizes and shapes which are distinctive and particularly attractive to buyers of art and gift goods who are looking for beautiful wares which register the out-of-the-ordinary note."

Although plain black flower blocks were also being made at this time, it wasn't until 1926 that black (without iridescence) went into general production at Diamond. Trade journal reports and advertisements mention console sets, vases, and liquor sets. These continued to be made until 1931, when the factory burned, with pieces from this period frequently found "with modernistic silver decoration" (*Crockery and Glass Journal*, July 1931). One example is the "Jack and the Beanstalk" decoration mentioned in the September 18, 1930, issue of *The Pottery, Glass & Brass Salesman*.

By 1931 Diamond was heavily into black glass as noted in this trade journal quote: "An extensive line of novelties in black glassware with silver decoration manufactured by the Diamond Glassware Company of Indiana, Pennsylvania, is being displayed...A notable offering is a line of jars, which are reproductions of classic Japanese temple jars. Also shown in the black glass treated with silver are wine and whiskey sets, including beverage glasses and decanters, also cake plates and other table pieces. Then, there is quite a line of lily bowls. The black glass is an unusually good color, while the shapes are attractive, the decanters being tall and attractively graceful. The silver work is well executed and, aside from conventional treatments, includes figures of roosters, dogs, and other animals..." *Pottery, Glass and Brass Salesman*, February 12, 1931.

Diamond Glass-Ware advertisement appearing in trade journal. Circa 1930s.

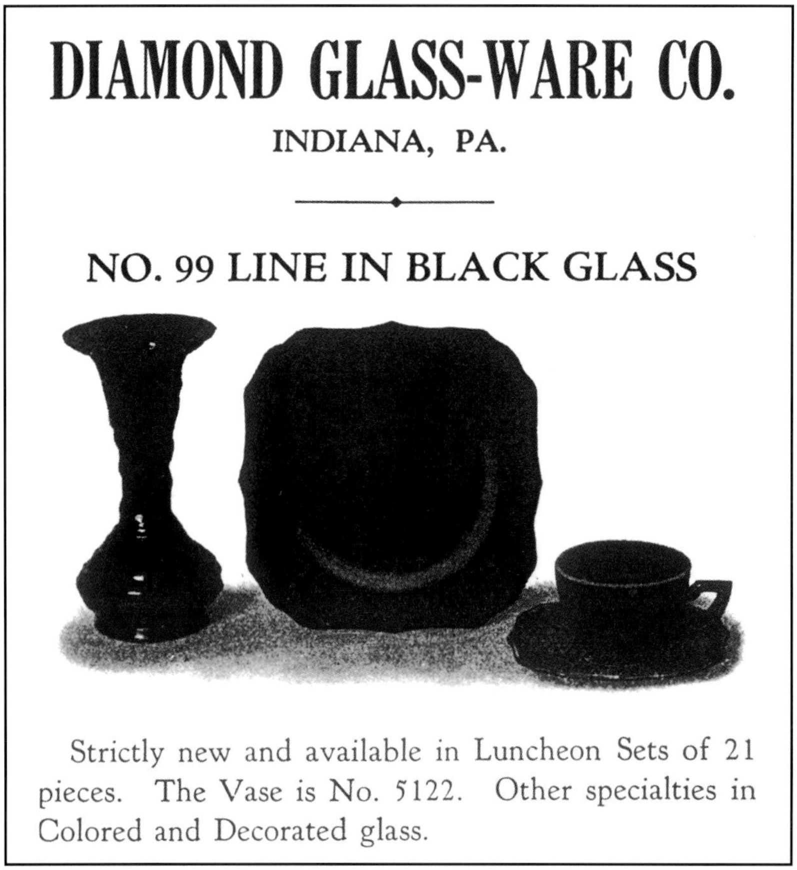

Diamond Glass-Ware advertisement appearing in trade journal. Circa 1930s.

Diamond Glass Co. advertisement appearing in *The Pottery, Glass & Brass Salesman*, September 18, 1930.

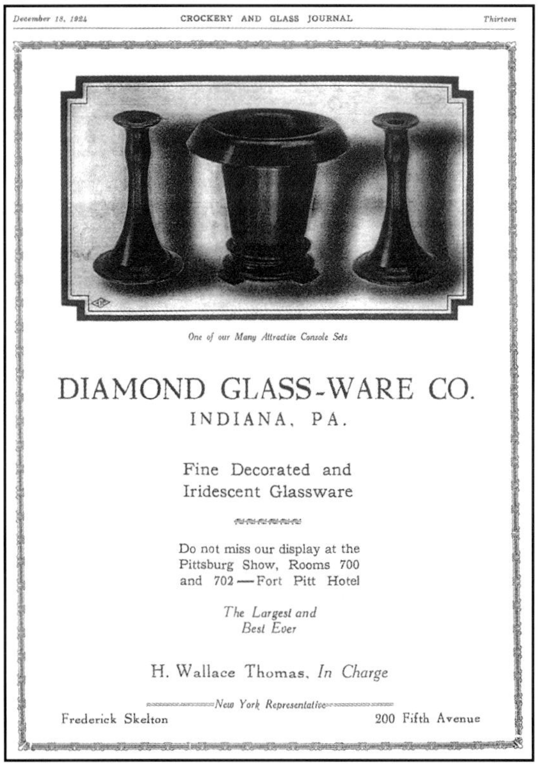

Diamond Glass Co. advertisement appearing in *Crockery and Glass Journal*, December 18, 1924.

Diamond Glass advertisement appearing in *Pottery, Glass & Brass Salesman*, February 16, 1928.

Dithridge & Co.

Pittsburgh, Pennsylvania
1827-1902

Dithridge & Company was founded in Pittsburgh, Pennsylvania, in 1827 as the Fort Pitt Glass Works. By 1867, the company was known as Dithridge and Son. Production in the earliest years included pressed lead glass known today as lacy glass. It then seems to have turned heavily to lamp chimneys after the mid-nineteenth century when kerosene lamps become popular. It wasn't until 1880, by which time the firm had been renamed Dithridge and Company that tumblers and bar goods were added to their cataloged line. In 1890 the Pittsburgh factory was abandoned and a new plant built in Jeannette, Pennsylvania. There Dithridge continued to operate until 1902 when they became a part of the Pittsburgh Lamp, Brass and Glass Company (a conglomerate that included the Pittsburgh Lamp and Brass Company and Kopp Lamp and Glass Company with Dithridge).

While lamps became the most significant part of their business, from 1890 on Dithridge also positioned itself as one of the major producers of opal (milk glass) novelties and, beginning at least as early as 1896, black glass. Trade journals do not mention this color again until 1900 when *Crockery and Glass Journal* reported: "Quite another style of ware, appropriately called 'Nubian,' is in black glass with embossed flowers in colors. It is striking and artistic, besides being the very newest novelty on the market. There are tobacco jars, ash trays and many dainty pieces for the dressing table and 'my lady's' desk."

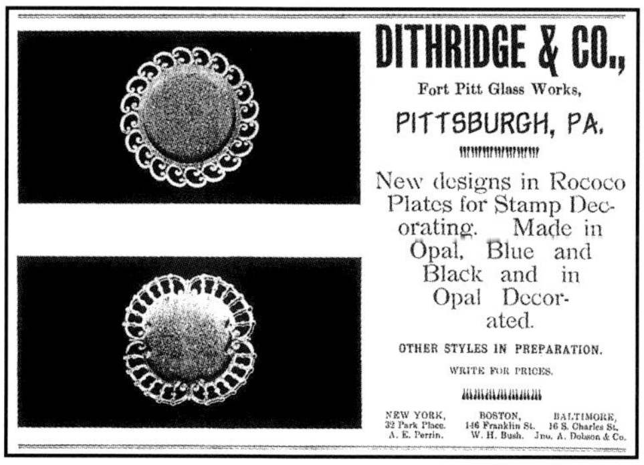

Dithridge & Co. advertisement appearing in *China, Glass and Lamps*, April 22, 1896.

Dunbar Glass Corporation

Dunbar, West Virginia
1923-1953

Dunbar Flint Glass Company existed as early as 1923 but produced exclusively lamp chimneys. The name change and new corporate charter of 6 July 1923 created the Dunbar Glass Corporation. Dunbar was created in the midst of a small industrial complex that had included several earlier glasshouses. Parts of some of those sites seem to have become part of Dunbar Glass over time.

Dunbar began to press glass in 1927, and a surviving catalog from 1931 includes several pages of black glass. Perhaps it was the scarcity of raw materials to color glass, the lack of labor, or the conversion to military production during World War II that ended black glass production at Dunbar. No indication of any black glass production at Dunbar existed by the time of or after WW II.

Later Dunbar production turned to lighting goods and in 1953 Dunbar withdrew entirely from pressed and blown glass to concentrate on drawn glass tubing. Shortly after that change Dunbar closed permanently.

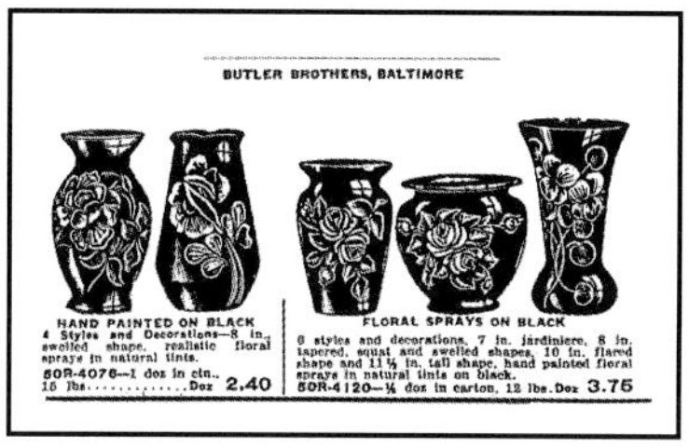

Black glass vases appearing in Butler Bros. catalog 1935. Note the second vase from the left is Dunbar's 498-132 from the catalog page. Butler Bros. claimed they shipped in the original manufacturer's "barrel" to insure quality packing, thus we can attribute the other shapes to Dunbar with considerable certainty.

Dunbar Glass Corporation original catalog page. 1931 Catalog.

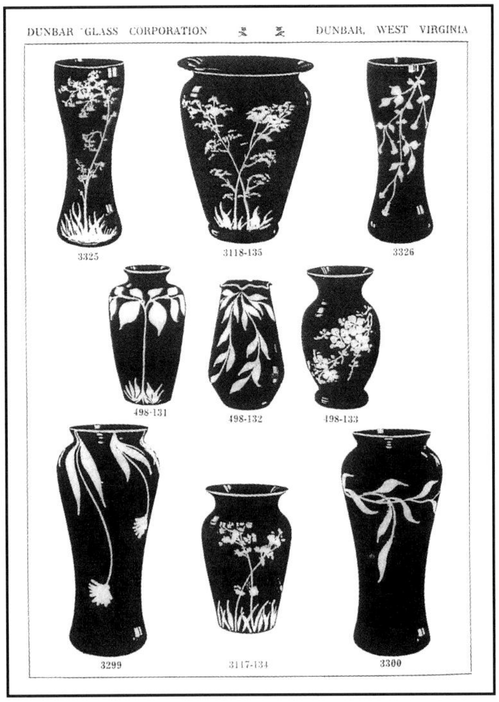

Dunbar Glass Corporation original catalog page. 1931 Catalog.

Dunbar Glass Corporation original catalog page. 1931 Catalog.

Duncan & Miller Glass Co.

Washington, Pennsylvania 1893-1955

Duncan & Miller began in 1893 as Geo. Duncan's Sons & Co. The company name was quickly changed to Duncan & Miller Glass Co. after John Miller, head mold maker, became a partner.

Initial production primarily focused on pressed tableware and barware. As tastes in glassware changed, Duncan & Miller's production also changed. Bowls, candlesticks, perfumes, figurines, and vases were all made, as were numerous other shapes. Blown glass was added to the existing pressed lines, and Duncan & Miller became known for its many etched and cut decorations on tableware and decorative accessories.

"For 1930 the Duncan & Miller Glass Co., of Washington, Pa., have added new items to some of their most popular lines. The particular feature of their new line is bowls, candlesticks, puff boxes and colognes in gleaming ebony glass, which is now at the crest of a vogue." *Crockery and Glass Journal*, February 1930.

Duncan & Miller was sold to the U.S. Glass Co. of Tiffin, Ohio, in 1955 and the assets, including molds, were moved to the Tiffin factory or the Glassport, Pennsylvania, factory of U.S. Glass. Some of the more popular and best selling Duncan & Miller lines were continued in production by U.S. Glass, but not usually in black glass.

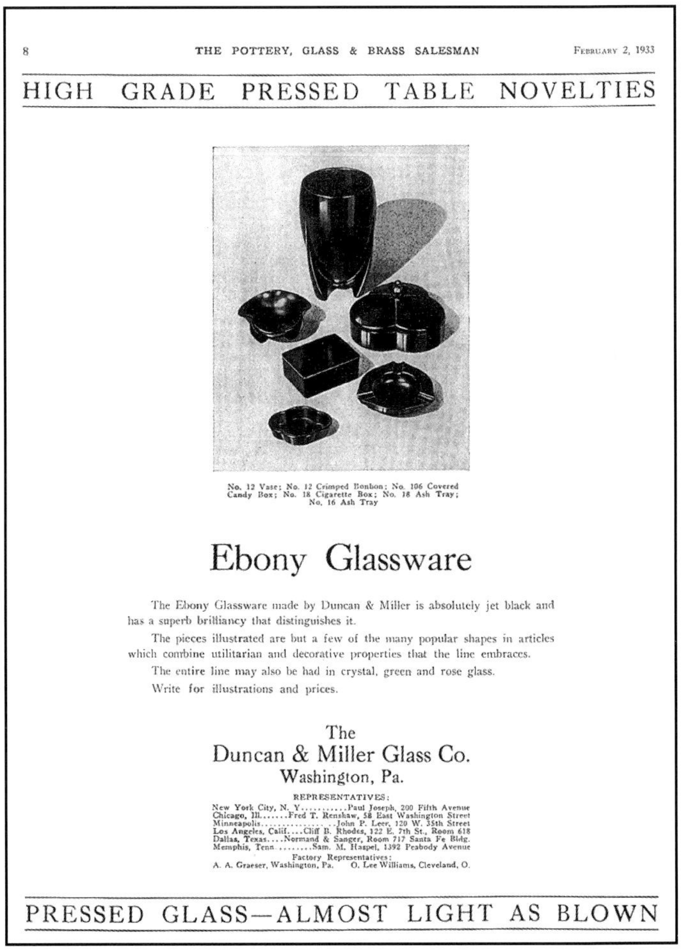
8 THE POTTERY, GLASS & BRASS SALESMAN February 2, 1933

HIGH GRADE PRESSED TABLE NOVELTIES

No. 12 Vase; No. 12 Crimped Bonbon; No. 106 Covered Candy Box; No. 18 Cigarette Box; No. 18 Ash Tray; No. 16 Ash Tray

Ebony Glassware

The Ebony Glassware made by Duncan & Miller is absolutely jet black and has a superb brilliancy that distinguishes it.

The pieces illustrated are but a few of the many popular shapes in articles which combine utilitarian and decorative properties that the line embraces.

The entire line may also be had in crystal, green and rose glass.

Write for illustrations and prices.

The
Duncan & Miller Glass Co.
Washington, Pa.

REPRESENTATIVES:
New York City, N. Y. .... Paul Joseph, 200 Fifth Avenue
Chicago, Ill. .... Fred T. Renshaw, 58 East Washington Street
Minneapolis .... John P. Leer, 120 W. 35th Street
Los Angeles, Calif. ... Cliff B. Rhodes, 122 E. 7th St., Room 618
Dallas, Texas .... Normand & Sanger, Room 717 Santa Fe Bldg.
Memphis, Tenn. .... Sam. M. Haspel, 1392 Peabody Avenue
Factory Representatives:
A. A. Graeser, Washington, Pa. O. Lee Williams, Cleveland, O.

PRESSED GLASS—ALMOST LIGHT AS BLOWN

Duncan & Miller Glass Co. advertisement appearing in *The Pottery, Glass & Brass Salesman*, February 2, 1933.

Duncan & Miller Glass Co. advertisement appearing in trade journal, 1933.

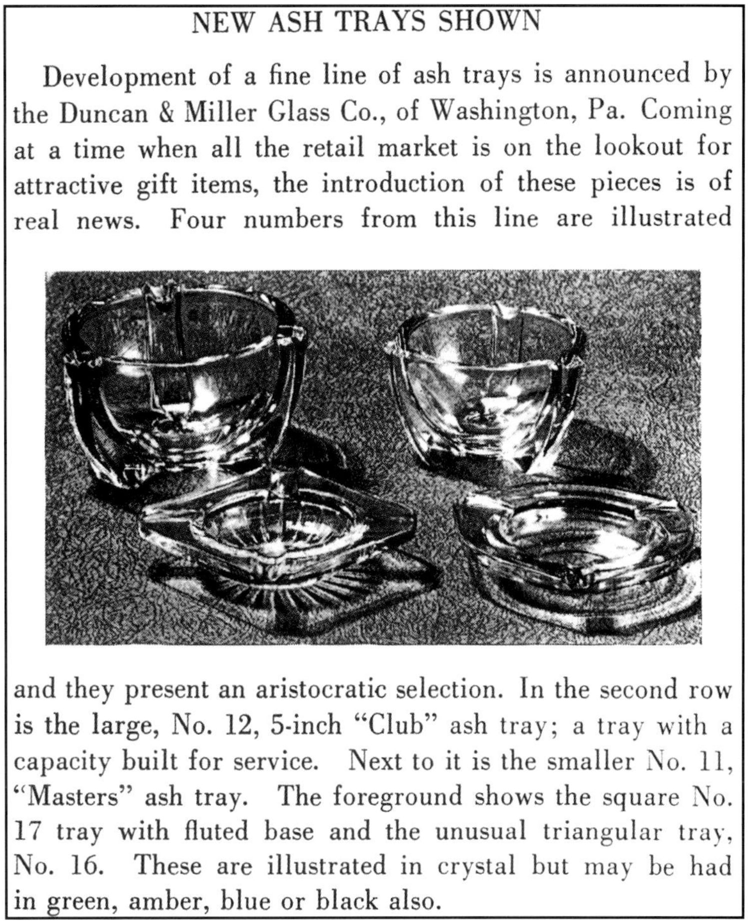

NEW ASH TRAYS SHOWN

Development of a fine line of ash trays is announced by the Duncan & Miller Glass Co., of Washington, Pa. Coming at a time when all the retail market is on the lookout for attractive gift items, the introduction of these pieces is of real news. Four numbers from this line are illustrated and they present an aristocratic selection. In the second row is the large, No. 12, 5-inch "Club" ash tray; a tray with a capacity built for service. Next to it is the smaller No. 11, "Masters" ash tray. The foreground shows the square No. 17 tray with fluted base and the unusual triangular tray, No. 16. These are illustrated in crystal but may be had in green, amber, blue or black also.

Duncan & Miller Glass Co. advertisement appearing in trade journal, circa 1933.

Duncan & Miller Glass Co. advertisement appearing in *The Crockery and Glass Journal*, circa 1930s.

Economy Glass and Economy Tumbler, see Morgantown Glass Works

Fenton Art Glass Co.

Williamstown, West Virginia
1905 - operating at the time of this writing

Fenton began in 1905 as a decorator using glass blanks produced by other companies. Fenton soon relocated to Williamstown, West Virginia, and built a new factory there. Fenton is credited with creating the iridescent ware we know today as Carnival Glass, and over the decades has utilized almost every color and glass treatment known. Black glass has figured prominently in their line at several periods over the century of production.

Black glass appears early in the Fenton offerings and reappears numerous times over the decades. Authors Margaret and Kenn Whitmyer note they found the first mention of black glass at Fenton in 1911. Black glass has figured prominently in Fenton production in recent years.

For several decades Fenton has coded their colors into the object's catalog number. It is not surprisingly that the contemporary Fenton code for black is BK. The following list, created by Debbie & Randy Coe for this book, is a check list of Fenton black glass made after 1968, including the years it appeared in the company catalogs. No black was produced in the years 1977-1980, 1983-1985, 1988-1989, 1995-1996, and 1998-1999.

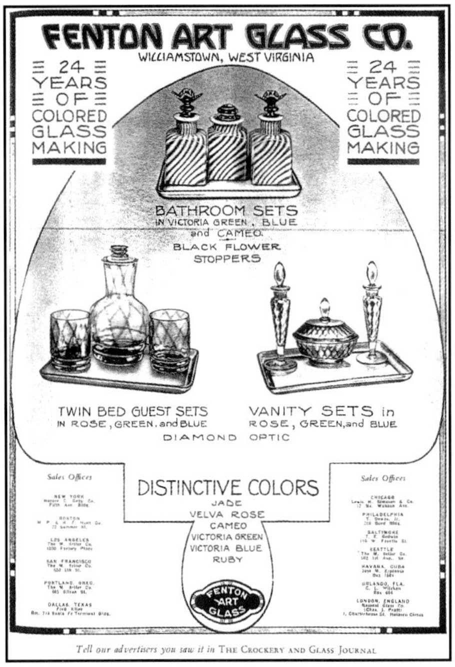

Left: Fenton Art Glass advertisement appearing in *The Crockery and Glass Journal*, March 1929.

Fenton Art Glass catalog page, 1969 catalog.

Fenton Art Glass cream and sugar appearing in Montgomery Ward catalog, Fall & Winter 1930-1931.

Candle Bowl 3872BK 1968 to 1974
Compote 4425BK 1968 to 1974
Compote 4429BK 1968 to 1975
Basket 8 1/2 inches 4438BK 1968 to 1975
Bud vase, tall 4453BK 1968 to 1975
Vase, footed 8 inch 4454BK 1968 to 1974
Bud vase 4455BK 1968 to 1976
Ashtray 6 1/2 inches 4469BK 1968 to 1969
Cigarette lighter 4495BK 1968 to 1969
Owl Decision Maker 5108BK 1968 to 1971
Vase, mandarin 8251BK 1969 only
Vase, empress 8252BK 1969 only
Boy and Girl praying figural 5100BK 1973
Temple Jar 7488BK 1981 and with Silver Poppies decoration in 1982
Bowl, 6 inches 7521BK 1981 and with Silver Poppies decoration in 1982
Ivy Ball 7522BK 1981
Vase 7550BK 1981 and with Silver Poppies decoration in 1982
Vase 7557BK 1981 and with Silver Poppies decoration in 1982
Vase, bud 7558BK 1981 and with Silver Poppies decoration in 1982
Vase 7561BK 10 1/2 inches 1981 and with Silver Poppies decoration in 1982
Temple Jar 7588BK 1981 and with Silver Poppies decoration in 1982
Vase 7561SX SX = Sophisticated Ladies, a sand carved decoration on black 1982
Vase, cylinder 7651SX SX = Sophisticated Ladies, a sand carved decoration on black 1982
Vase, sphere 7655SX SX = Sophisticated Ladies, a sand carved decoration on black 1982

The following is a list of the decorated black glass in the line. Often you could find all the pieces from the decorated line in the Fenton Gift Shop unpainted. It seems possible that for as many decorated items as there are, there are potentially almost as many of those same items not decorated. The undecorated pieces did not appear in the catalogs. Once you figure out what is in the catalog line, then it makes it easy to decide what year an item was made if it is found undecorated.

1986 to 1987 Silhouette - black glass decorated with a Japanese motif
Vase, oval 12 inches 8802EK
Vase, sphere 9 1/2 inches 8804EK
Vase 7 1/2 inches 8806EK
Vase 10 1/2 inches 8812EK

1990 to 1993 Copper Rose - black glass decorated with pink roses and copper accents and factory coded as KP.
Bear 5151KP 1990-1993
Cat 5165KP 1990-1993
Happiness Bird 5197KP 1990-1993
Box, Heart 5780KP 1990-1993
Box, Wavecrest 6080KP 1990-1993
Compote 6322KP 1990-1993
Bell, Paisley 6761KP 1990-1993
Vase 6 inches 7348KP 1990-1993
Bowl, rolled edge 9 1/2 inches 7523KP 1990-1993
Bell, aurora 7630KP 1990-1993
Vase 7 1/2 inches 7696KP 1990-1993
Clock 8691KP 1990-1993
Vase 8 1/2 inches 8817KP 1990-1993
Bell, Drape 9266KP 1990-1993
Slipper, Rose 9295KP 1990-1993
Candleholder 9372KP 1990-1993
Atomizer 7948KP 1992
Clock, mini 8698KP 1992

1994 black glass decorated Autumn Leaves on Black, factory code AW:
Vase 9 1/2 inches 2756AW 1994
Vase 5 inches 2757AW 1994
Vase 8 inches 2758AW 1994
Rose Bowl 3 1/2 inches 2759AW 1994
Fox 5226AW 1994
Unicorn 5253AW (the only piece to also be made for the catalog in plain black 5253BK) 1994
Basket 10 1/2 inches 6572AW 1994
Candy 9 1/2 inches 7380AW 1994
Clock 6 inches 8600AW 1994
Candleholders 8 1/2 inches 9021AW 1994
Slipper, Rose 9295AW 1994
Bell, Aurora 9667AW 1994

1997 the Medallion Collection appears in the Fenton catalog. Only the following three items in the Medallion Collection were black glass and hand painted; other items from this collection had other base colors of glass.
Cat 5165X7 1997
Owl 5258Y8 1997
Vase 8 inches 7565X5 1997

Fenton round logo 8599E3 2001
Slipper 9295J2 flowers on black slipper - in the slipper collection 2001
Egg 5146J2 fish on black egg - in the egg collection 2001
Rooster Box 8 1/2 inches 4680OA Folk Art Animals 2001
Rooster, standing 8 1/2 inches 5257OA Folk Art Animals 2001
Rooster 2 3/4 inches 5265OA Folk Art Animals 2001
Rooster 5 1/2 inches 5292OA Folk Art Animals 2001
Rooster box 4 1/2 inches 6483OA 2002 w line of Folk Art Animals
Rooster box 8 1/2 inches 4680B3 2002
Rooster standing 8 1/2 inches 5257B3 2002
Rooster 2 3/4 inches 5265B3 2002
Rooster 5 1/2 inches 5292B3 2002
Rooster box 4 1/2 inches 6483B3 2002
Egg 54146Z8 2002
Vase, square 9866BY Tiger Lilies on Black - Family Signature line 2002
Witch, Halloween 5284H6 2002
Cat, Halloween 5291HN 2002
Basket, Halloween 9 inches 5936H9 2003
Box, covered 71083Y floral on black - Designer series 2003
8970DJ Rainforest on Black - Centennial Collection 2003
Witch, Halloween 5284Y2 (same figures as prior year just new decoration) 2003
Cat, Halloween 5291Y4 (same figures as prior year just new decoration) 2003
Basket, Halloween 9 inches 5936ZT (same figures as prior year just new decoration) 2003
Basket, Halloween 10 1/2 inches 3097ZB 2004
Owl, Halloween 5258ZB 2004
Ghost, Halloween 5278YB (same figures as prior year just new decorations) 2004
Witch, Halloween 5284ZB (same figures as prior year just new decorations) 2004
Cat, Halloween 5291ZB (same figures as prior year just new decorations) 2004
Tree, Halloween 5563ZB (same figures as prior year just new decorations) 2004

To learn more about contemporary production at Fenton visit Fenton on line at www.fentonartglass.com

Fostoria Glass Co.

Moundsville, West Virginia
1887-1984

Fostoria Glass Co. opened in 1887 in the town of Fostoria, Ohio, from which the glass company took its name. Due to shortages in the natural gas supply, the company relocated to Moundsville, West Virginia, in 1891.

Fostoria produced extensive lines of pressed tableware patterns, their most popular being American, arguably the longest produced glass tableware pattern in the United States. As with most other successful glass companies, changes in consumer's taste forced changes in production in the nineteen teens and twenties. Blown wares, cuttings, and plate etchings were added to the line and endured for many decades.

Black was introduced by Fostoria in 1930. "Striking Ensembles of Fostoria Glass," reports a trade journal of that era. "Entering the main showroom, recently, one beheld on the right a table set forth with extreme formality in a service of the new Crystal Night pattern: all-black flatware, black footed stemware with plain crystal bowls, black centerpiece and comports and candlesticks decorated with bands of gold ... a rich, slightly somber, yet ultra-smart assemblage. Beyond it, a table set for bridge, with the Saturn pattern, so-called by reason of the concentric rings of black enamel around the bowls of the black footed stemware, accompanied by square, scalloped-edge crystal plates with three black enamel bands." *Crockery & Glass Journal,* March, 1930.

Fostoria closed the factory in 1984. The name and other assets were acquired by Lancaster Colony and continue to be used to this day around the world at the time of this writing.

Fostoria Glass catalog page noting items available in Ebony. 1927.

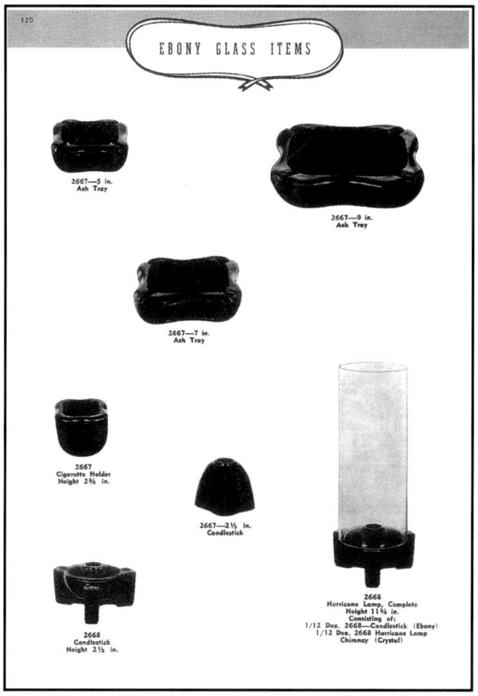

Fostoria Glass original Ebony Glass catalog page dated 1955.

Left:
Fostoria Glass advertisement appearing in *China, Glass and Decorative Accessories*, August 1953.

Right:
Fostoria Glass original Ebony Glass catalog page dated 1955.

118

EBONY GLASS ITEMS

2453
Lustre
8 U. Drop Prisms
Height 7½ in.

2467—7½ in.
Vase

2538
Place Card Holder
Height 2¾ in.

2567—7½ in.
Footed Vase

2545—2 in.
"Flame" Candlestick

2545
Candle Lamp, Complete
Consisting of:
1/12 Dz. 26 Candle Lamp Base (Crystal)
1/12 Dz. 26 Candle Lamp Chimney (Crystal)
1/12 Dz. 2545 2 in. Candlestick (Ebony)

2618
Cigarette Box and Cover
Length 5½ in. Width 4¼ in.

2592
Cigarette Box and Cover
Length 6 in. Width 3¼ in.

2592
Oblong Ash Tray
Length 3¾ in. Width 2¾ in.

FOSTORIA GLASS COMPANY, MOUNDSVILLE, WEST VIRGINIA — 1958

Fostoria Glass original Ebony Glass catalog page dated 1958.

Fostoria Glass original consumer booklet. Dated 1981.

117

EBONY GLASS ITEMS

2288—8½ in.
Tut Vase, Handled

2402—9 in.
Bowl

2402—2 in.
Candlestick

2404—6 in.
Vase

2409—7½ in.
Vase

2428—13 in.
Vase (ill.)
2428—9 in.
Vase
2428—6 in.
Vase

2428—7 in.
Round Bowl

2430
Candy Jar and Cover
Height 5¾ in.

2430—2 in.
Candlestick

2430—11 in.
Bowl

FOSTORIA GLASS COMPANY, MOUNDSVILLE, WEST VIRGINIA — 1958

Fostoria Glass original Ebony Glass catalog page dated 1958.

Fostoria Glass cover to original consumer booklet. Dated 1981.

Fostoria Glass original consumer booklet. Dated 1981.

H.C. Fry Glass Company

Rochester, Pennsylvania
1902-1933

Opening in 1902 in Rochester, Pennsylvania, as the Rochester Glass Company, the name was soon changed to the H.C. Fry Glass Company. Initial production consisted of high quality pressed and blown tumblers, but this was soon expanded to a full line of blanks for cutting by Fry and other decorating companies. Fry's reputation for the exceptional quality of their blanks and their cuttings remains highly regarded today.

Although cut glass remained in the Fry line into the 1920s, the market for such high end ware had diminished. In 1915 Fry introduced a line of heat-resistant ovenware, both in crystal and pearl (or opalescent). In 1922 Foval, a line of art glass in pearl with Delft blue, jade green or rose trim, was introduced. While these lines were popular, they were not enough to keep Fry from being forced into receivership in 1925. They remained in operation, introducing a variety of transparent colors, and were reorganized in 1929. The new management was optimistic, but was unable to withstand the pressures of the Depression years, finally closing the company in 1933.

One of the final colors introduced around the time of the reorganization in 1929 was black, often used in combination with crystal or color. "The Fry presentation of amber and black is...receiving an enthusiastic reception by the trade. A combination of rock crystal bowl and ebony foot is the basis of another pleasing Fry offering." The same report in the November 1929 *Crockery and Glass Journal* went on to mention a "line of unusual vases newly presented by the H. C. Fry Co. An ebony vase with gold webbing, and an ebony vase with a gold and silver webbing are the most striking."

Surely one of the most interesting creations using black glass from Fry would be found in this mention: "A novel and clever line of 'Lighthouse radio Lamps' is contributed by the H. C. Fry Glass Company and has met with considerable success at the hands of the local retail trade. The lamps are manufactured entirely of glass. Each lamp consists of two pieces- a black base in the shape of a lighthouse and a large bubble in crystal and several pastel tints. The light bulb is fitted inside the base and casts its glow upward through the ball. The effect is at once unique and pleasing." *Pottery, Glass, and Brass Salesman,* July 16, 1931.

Noteworthy in black from Fry are several combination pieces utilizing black glass with other colors. An impressive selection of stemware with black feet and/or stems as well as elaborate high-end art glass can be found.

Fry Glass "Diamond Optic Reeded in Black" original catalog page circa 1930.

Hazel-Atlas

Wheeling, West Virginia, and factories including Clarksburg, West Virginia
1902-1956

With origins in a glass factory in Wellsburg, the Brady Brothers, owners of the company, merged their interests in the Hazel Glass Company and the Atlas Closure Company to form what would grow to be a nationwide glass empire. The merger was accomplished in 1902 and Hazel-Atlas was formed.

The Republic Glass Co. (1900-1901) in Clarksburg, West Virginia, became part of the Hazel Atlas Company and was for many decades the dominant Hazel-Atlas tableware producing factory. Headquarters and corporate offices remained in Wheeling, West Virginia. Major plants were located in Washington, Pennsylvania, and up to thirteen other cities across the U.S. It has proven difficult to attribute most Hazel-Atlas products to a specific factory, but much of the tableware and black glass was shipped from the Clarksburg site per the company catalogs. This suggests production of those lines at the Clarksburg location, while several of the other sites made bottles, jars, packing tumblers, and diverse items in milk glass, cobalt blue, amber, and much, much more.

The Hazel-Atlas empire was sold to diverse interests in 1956, with the Clarksburg site becoming a division of Continental Can Company.

A.H. Heisey & Co.

Newark, Ohio
1896-1957

A.H. Heisey & Co. was established in Newark in 1896 to make pressed glass tableware and barware. By the late nineteen teens, blown ware, acid plate and needle etchings, and cuttings were added to the line. Heisey was famous for its cleverly designed animals by Royal Hickman, which were produced in the 1940s.

Black glass production was done only on an experimental basis and finding pieces on the market is highly unlikely. When the factory closed in 1957 six black Lariat pattern plates were found. A few pieces have been seen such as marmalade jar bases, a colonial oval plate and small plates in the Town and Country line. The probable date of the experimental black glass was the mid-1950s when others in the market place shared with Heisey were also producing black glass. Heisey's black formula was derived from a formula using cobalt, thus the glass appears a dark, deep blue under strong light not the usual violet or amethyst.

Heisey ceased business in December 1957 and its assets, including molds, were sold to the Imperial Glass Corporation of Bellaire, Ohio, who continued some of the more popular lines and patterns.

J.L. Houze Glass Company

Point Marion, Pennsylvania
1914 - operating at the time of this writing

Opening in 1902 as Federated Glass Company to make flat window glass, the company became L.J. Houze Convex Glass Company in 1914. Specializing in "convex" glass used in military application during World War I, the company was producing slag or marbled glass by the 1920s. Ashtrays, flowerpots, small novelties, and private mould work for countless other manufacturers grew the company. By the early 1950s Houze perfected a process to transfer photographs and artwork using silkscreen printing to permanently fired decorations on glass. From glass greeting cards marketed in ready to mail boxes to multitudes of commemorative and souvenir ashtrays, the application on dark gray and black glass was popular for several years.

The late 1960s, we were told by authors Felt and Stoer, in their candlestick book series, saw Houze turning to decorating rather than manufacturing glass. They remain in the business of creating specialty items at the time of this writing.

Imperial Glass Corporation

Bellaire, Ohio
1904-1984

Imperial Glass produced primarily pressed glass tableware, barware, and other items from its earliest days. For a short time in the 1920s it produced a line of handmade art glass similar to that made at Durand, Fenton, and others.

Imperial was a large company and over time acquired the assets of many of its once competitors, including Central Glass Co., A.H. Heisey & Co., and Cambridge Glass Co. Popular patterns at Imperial, including Candlewick and Cape Cod, sold in great quantities for long periods. These patterns were part of the equation that helped Imperial survive well beyond the time of many other handmade glass houses. Imperial produced black glass in the 1930s and again in the1950s.

Imperial Glass closed in 1984.

Imperial Glass catalog page circa 1930.

Imperial Glass Black and Gold. Original page from catalog "53" circa 1953.

Imperial Glass original Black Suede or satin glass. Original page from catalog "53" circa 1953.

Indiana Glass Company

Dunkirk, Indiana
1904-2002

The Indiana Glass Company was founded in 1904 when the Beatty-Brady Glass Works in Dunkirk, Indiana, was leased from its owner, The National Glass Company, by Frank W. Merry, the site's previous superintendent. After National failed, the factory was purchased and Merry remained president until his death in 1930. The production from the company varied greatly, including both machine-made and handmade glassware, ranging from Sandwich reproductions and popular Depression glass patterns to automobile headlight lenses.

In the 1950s, the company underwent a period of financial difficulty that culminated with their becoming a division of the Lancaster Colony Corporation. Even before this occurred, Indiana provided much of the glassware sold by Pitman-Dreitzer & Company's division, Colony Glassware. That relationship continued when Pitman-Dreitzer also became a part of the conglomerate Lancaster Colony Corporation in 1962. Over the years molds were acquired from many other companies including McKee Glass, U.S. Glass, Duncan & Miller, Paden City, Jenkins, Imperial, and Fostoria. In 1970 Tiara Exclusives was formed to offer sales through a home party plan and new life was given many of the Indiana patterns as well as other patterns that had been acquired by them. Most of the Tiara production emanated from Indiana's Dunkirk factory. When handmade operations ceased in the mid-1980s, Tiara began selling glass from other companies. While the Indiana label is still found today, the production at the Dunkirk plant was discontinued in late 2002 and consolidated in Lancaster Colony's Sapulpa, Oklahoma, factory.

There was some Indiana production of colored glass in the 1920s and 1930s and this appears to have included black, but the earliest confirmed date for black glass is 1973 when Indiana's Pyramid pattern was reissued for Tiara's Art Deco Collection. "Tiara has dusted off the old moulds, dating from World War I, and given them new life with its black patent-leather glass. The crisp wet look is as modern as tomorrow, yet coordinates beautifully with your crystal or colored ware – gold, carnival or horizon blue" the catalog announced. A number of other patterns followed, with black continuing to be made into the 1990s, both for Tiara and also appearing in Indiana's own catalog.

Tiara Glass original catalog page 1978 catalog.

raftsmenship and beauty make these gifts everyday favorites . . . . .

A

B

C

D

(A) 10419 Monarch 8½" Bowl - This elegant bowl enhances flowers and fruits.
(B) 10063 Kahlua Mug, Pair - Palm trees and warm breezes come to mind as your guests sip their mai tais or other cool drinks from these 6-1/8" tall mugs.
(C) 10422 Monarch Large 47 oz. Pitcher - Craftsmenship at its best makes this impressive 8" tall pitcher a work of art.
(D) 10424 Monarch 8 Oz. Goblets, 4 Pcs. - The beautiful deep cut pattern of these 6¼" goblets sparkles in any light.

Tiara Glass original catalog page 1978 catalog.

sparkling color complements...

A

B

C

D

Brighten any room with these colorful and functional gifts. An array of colors that complement any decor. Perfect mood setters, accent pieces, or memorable gifts.
(A) 10026 Lotus Ash Tray - The beautiful floral design of this tray is appreciated by smokers and non-smokers alike. The lovely Horizon Blue tray is 9½" in diameter.
(B) 10023 Crystal Eagle Ash Tray - The exceptional art work of this massive tray makes it a favorite of collectors of Early Americana. The Crystal tray blends with any decor. It is 10½" in diameter.
(C) 10022 Gold Eagle Ash Tray - Woodtones are complemented by the rich golden hues of the Eagle Ash Tray. Perfect for the executive desk. The beautiful eagle tray is 10½" in diameter.
(D) 10001 Smoker Set, 6 Pc. (Lighter, Cigarette Urn, 4 Ash Trays) - This elegant set graces any room or office. A practical gift for that special someone.

24

Tiara Glass original catalog page 1978 catalog.

Jefferson Glass Company

Steubenville, Ohio, and Follansbee, West Virginia
1900 - circa 1933

Jefferson Glass opened in Steubenville, Ohio, in 1900 where they remained until 1906. A new factory was constructed and the company relocated to Follansbee, West Virginia in 1906. Wares were handmade pressed glass including the popular lines of Chippendale glass from 1908 until 1919. In 1910 lighting was introduced into the Jefferson line and by 1920 Jefferson had turned exclusively to the production of lighting goods. A number of small glass electric bedroom and tablelamps were made by Jefferson in the 1920s and '30s. Some can be found with paper labels still intact on the underside of the base.

The factory was sold at auction in 1933.

Lancaster Glass Company

Lancaster, Ohio
1908-1933

The Lancaster Glass Company was located in Lancaster, Ohio, where it was put into operation in 1908 by Lucian B. Martin, the founder of the Fostoria Glass Company. He was also previously associated with the Ohio Flint Glass Company and the Hocking Glass Company – both of Lancaster, Ohio. In 1923, in a demonstration of how small the world of the glass industry seems to be, the Lancaster Glass Company was purchased by the Hocking Glass Company and operated as a division of Hocking until 1933 when both companies were absorbed into the Anchor Hocking Glass Corporation.

Where Hocking concentrated on producing machine-made glassware, Lancaster continued in the manufacture of handmade wares and also specialized in decoration on glass. A third factory belonging to the same family, the Standard Glass Manufacturing Company, also did decorations (mostly etchings and light cuttings) on blanks produced by Lancaster and Hocking.

Black was the first color offered by Lancaster. In April 1916, the *Crockery & Glass Journal* reported a Boston representative had "...received samples of new decorated candlesticks in etched color effects, new birds and a black cemetery vase from the Lancaster Glass Co."

The Pottery, Glass, & Brass Salesman announced the debut in March 16, 1916: "The Lancaster Glass Company, of Lancaster, Ohio, has brought out a large line of glassware which they are pleased to call 'Florentine Jet.' The ware is of a rich, opaque glass, and is obtainable plain and in a variety of attractive decorations. The latter include floral and other designs in enamel colors. Various styles of flower holders that are in vogue just now are made in 'Florentine Jet' glass, including flower bowls in many shapes and sizes, perforated flower blocks, handled baskets, vases, etc. There are also several attractive items for other uses, including candlesticks, trays, etc. Among the decorative treatments is a particularly pleasing pansy design in bright colors, likewise a fruit decoration, and several border effects in white."

Almost fifteen years later found the same trade journal describing Lancaster's newest pattern in its December 25, 1930, issue as "having something of the classic about it and at the same time it is absolutely up to date." They went on to note, "it is interesting that this line, like other Lancaster offerings, is now obtainable in black as well as in pink or green glass. The Lancaster concern intends to make quite a feature of its black glass next year. Many items will be shown entirely in black and others will show black trimmings. The black glass, of course, is a pot product and is a splendid jet black tone."

The Salesman mentioned a new grouping of black glass vases in January 14, 1932, "...with hand-painted decorations in silver and color," and made a special notice of a new salt and pepper set in March 3, 1932 "... in an opaque finish. The opaque finish, it was decided, was more practical for kitchen use because of the ease with which the usual crystal sets are soiled. The new sets come in a jet-black shade."

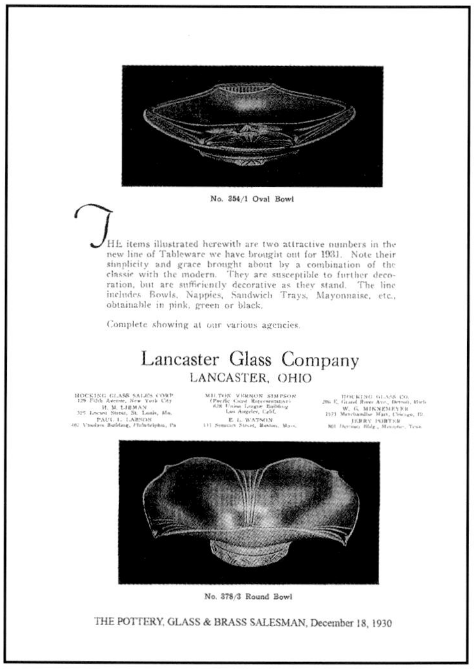

Lancaster Glass advertisement appearing in *The Pottery, Glass and Brass Salesman*, December 18, 1930.

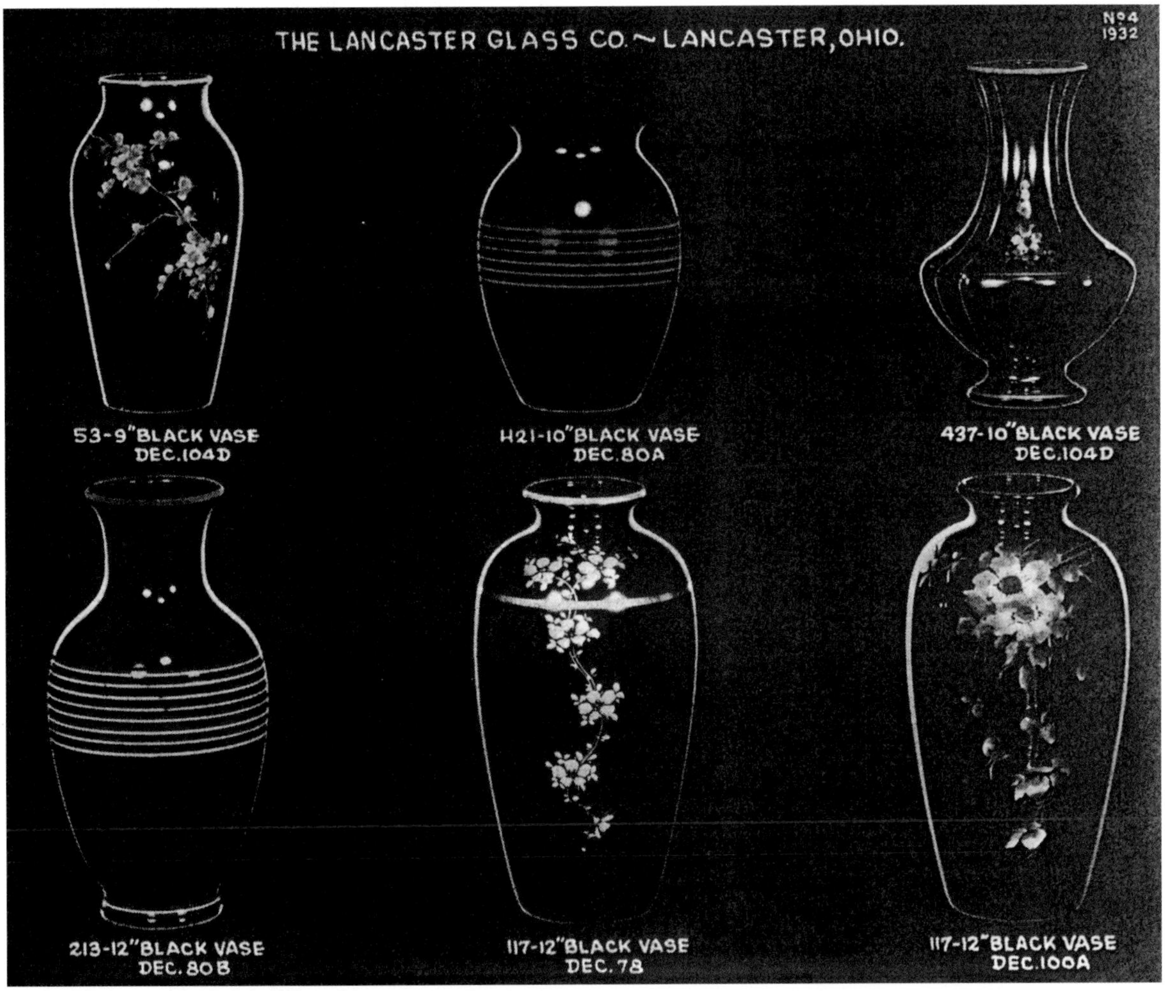

Lancaster Glass original catalog page dated 1932.

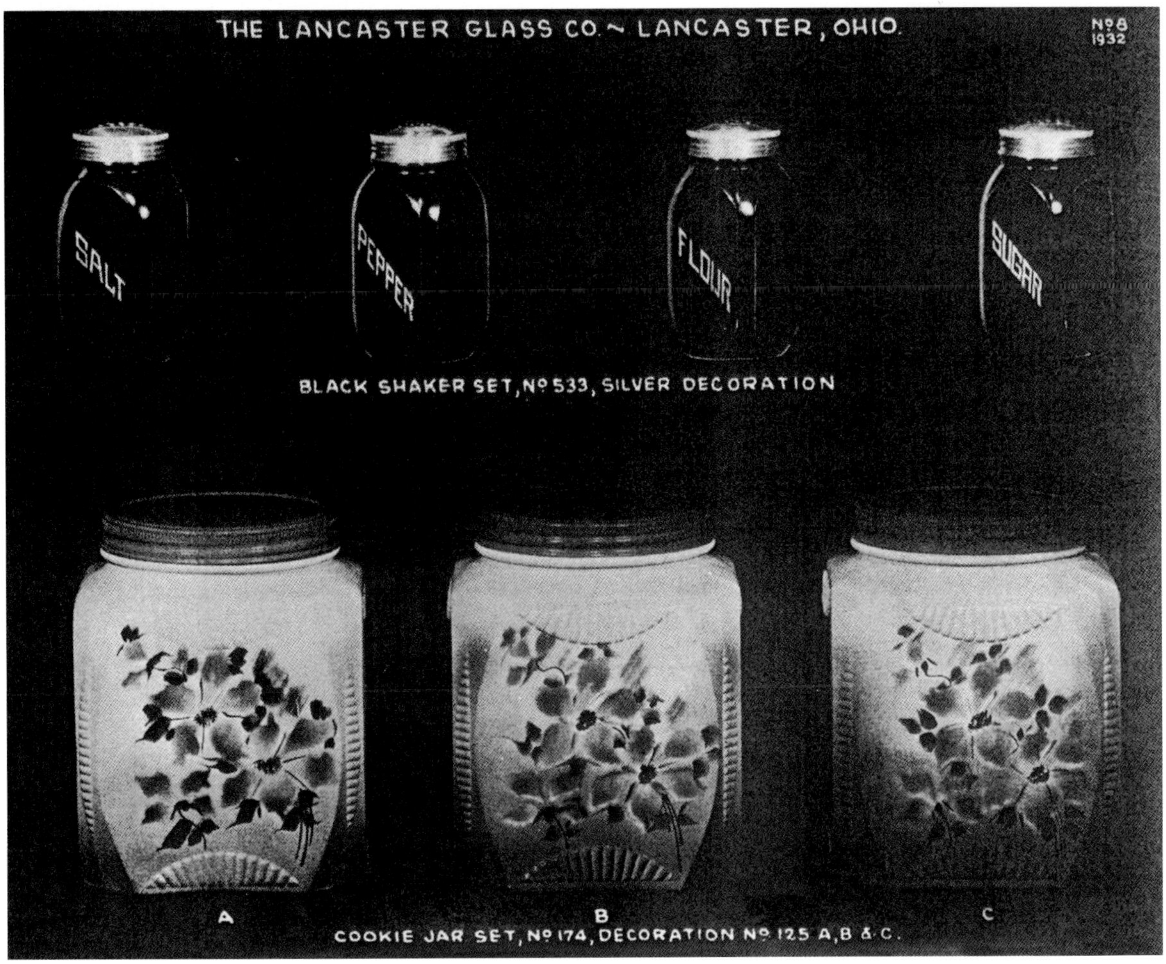

Lancaster Glass original catalog page dated 1932.

Lenox Crystal

Mt. Pleasant, Pennsylvania
1966 - closed U.S. factory 2003

Lenox purchased the plant of Bryce Bros. in Mt. Pleasant, Pennsylvania, in 1966. It was predominately a handmade stemware factory. The peak of black production was in the mid-1970s to coordinate with the many Lenox china patterns produced by the parent company. By the early 1980s, black was fading from Lenox stemware production. By the mid-1980s, black was probably no longer in production at Lenox.

Libbey Glass Company

Toledo, Ohio
1888 - still operating at the time of this writing

The Libbey Glass Company traces its origins back to 1818 and the New England Glass Company of East Cambridge, Massachusetts. Edward Drummond Libbey moved the company to Toledo, Ohio, in 1888, having changed the name to W. L. Libbey & Son Company in April of that year.

In the early years Libbey became well known for its rich cut glass and also for art glass, including lines called Amberina, Pomona, and others. The company took advantage of the 1893 World's Fair to set up a glass-making exhibit, making Libbey a household name because of the exposure and press generated. Following the decline in popularity of rich cut glass, the company struggled.

In a 1931 attempt to recapture the high-end market in glass, J.D. Robinson, then president of Libbey, hired A. Douglas Nash to develop a series of designs for Libbey. Nash had operated the Tiffany glass factory until it closed in 1930. What was to become known as the Libbey-Nash series debuted in 1933 with a sumptuous catalog. In the midst of the depression, the Nash designs failed after two short years. This was the period in which Libbey indulged in the production of designs that included black glass. It was a short period of production indeed.

Libbey Glass original catalog 1933. Libbey–Nash line.

Libbey Glass original catalog 1933. Libbey–Nash line.

The known Nash designs appearing in black glass include:

K-8, the Silhouette pattern of stemware and accessories. The animal stems in this line are found in black with crystal bowls.

K-5 is a set of bookends with a very angular, blocky art deco design.

K-522 are candlesticks in a tall, one light form with five octagonal black disks separated by crystal column.

Libbey, long known as a producer of high end hand cut crystal, produced a line of black over crystal where the black glass was cut away to reveal the underlying crystal. This "black cut to clear" dates to roughly the same period as the Nash pieces and was indeed a high end, expensive product when produced.

Libbey Glass original catalog 1933. Libbey–Nash line.

Libbey company catalog 1986.

Liberty Works

Egg Harbor, New Jersey
1903 - 1931

Liberty Cut Glass Works opened in 1903 as a decorating company doing cutting in the Rich Cut style on heavy glass blanks from other companies. The name was changed in 1917 to reflect their diversification and the firm became simply Liberty Works. Liberty began their own hot glass production in circa 1923. Their wares continued to include hand cut glass but in the lighter, gray cut manner. Glass manufactured by Liberty included pink, green, light blue, wisteria, cobalt blue, and black predominately in the forms and styles today known as depression glass.

Liberty fell victim to the economic hard times of the Great Depression and closed circa 1932.

Louie Glass Co.

Weston, West Virginia
1926-1996

The complex interconnected ownership of three Weston glasshouses, Louie Glass, Weston Glass, and West Virginia Glass Specialty, make certain attribution of many forms and lines common to Louie difficult when seeking to tie a specific piece or pattern to a single company. Molds, workers, and orders were shipped about at various times to fill orders. Local lore has it that following a June 1940 labor strike no colored glass was produced in the Weston factories on a large scale again.

Various early Weston factories may have produced green bodies with black feet and crystal bodies with black feet in stemware lines. Pitchers had black handles and lid on green or colorless bodies as well. Louie Glass and the allied factories were mouth blown or handmade glassware. Pieces were rarely pressed here but blown into molds.

Louie ceased to exist in 1996 when it assumed the name of its then parent company, Princess House.

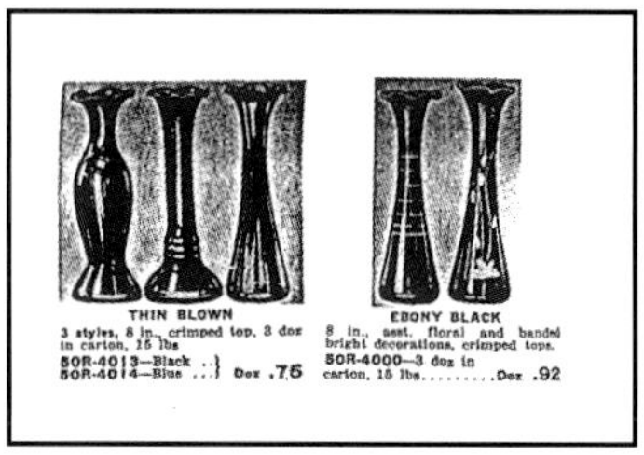

Louie black glass "thin blown" and "Ebony" vases offered in Butler Brothers 1935 catalog.

Louie black based tumblers with known Weston area cutting as offered with what appears to be Imperial Glass Molly pattern plates, etc. Montgomery Ward, Fall & Winter 1930-1931 catalog.

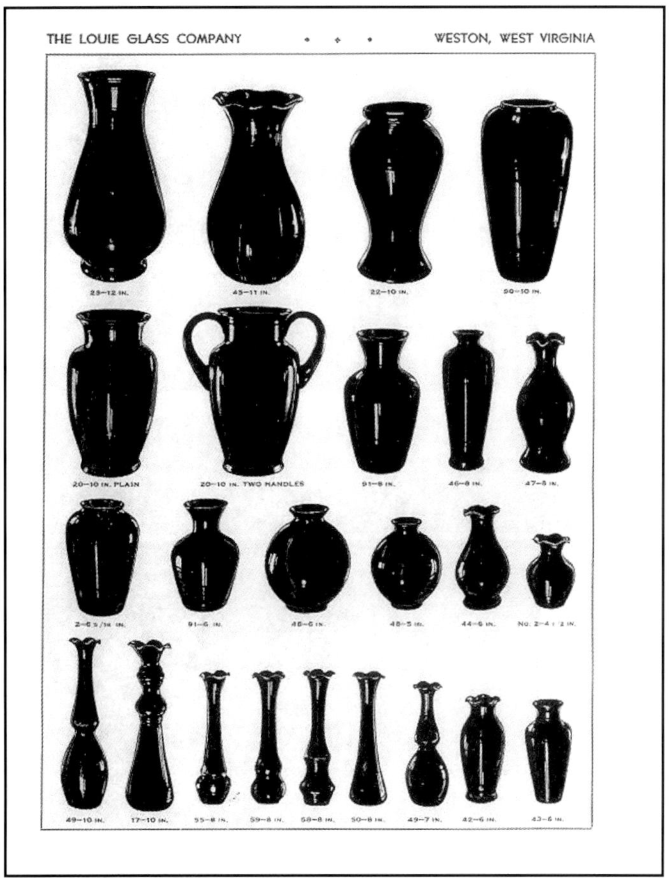

Louie Glass original company catalog page. 1930s.

Maryland Glass Co.

Cumberland, Maryland
1918-1941

Maryland Glass was located in Cumberland, Maryland, and operated there from 1918 until 1941. It was active as a glass decorator but also produced its own hot glass. It is reported Maryland produced lines of Hokey-pokey stemware (hand blown bowls attached to pressed glass stems), bud vases, and a limited number of other shapes in black glass. Many of the Maryland Glass vases are forms similar to, and often identical to, those made by Central, the Weston companies, and others in black glass.

McKee Glass Company

Pittsburgh and Jeannette Pennsylvania
1854-1961

The McKee family was associated with glassmaking as early as 1824. With such a long history, the company name has changed numerous times during its long history, particularly as partners came and went and the finances of the company at times grew and at times dwindled. Originally the company made window glass and bottles.

Pressed tableware was primarily made after the company was established in 1854 under the then name of McKee Brothers. In 1888 their factory moved to the new town of Jeannette, Pennsylvania, named for the wife of Henry Sellers McKee, who created the new town for workers in his new factory. Pieces during this period may bear the impressed mark Prescut.

In 1899 the firm joined the National Glass Co., a combine of several companies. The combine struggled from its early years although the McKee plant continued to be successful. A new company named McKee-Jeannette Glass Co. leased the plant from National in 1904 and by 1908 was founded as an independent company, the McKee Glass Co. The failure of National Glass in 1908 also wrecked McKee-Jeannette Glass. It re-emerged in 1910 as McKee Glass and was successful under that name for decades.

As early as 1923 McKee was producing some black glass, "Some of the colored bowls making a striking appearance on black glass bases" states *China, Glass & Lamps* in reference to McKee in a January 15, 1923, story.

In 1929 *The Pottery, Glass and Brass Salesman* noted a "new Line of Black Glassware from the McKee Glass Company." The accompanying article reads, "Kelly & Reasner, Shops Building, 17 North Wabash Avenue, the local agents for McKee Glass Company, Jeannette, Pa., are showing some of the new black glassware recently brought out by this factory. All of the articles in the line are good to look at, being graceful in shape. Stemware has the fluted effect. One item not always included among colored glassware which is available in this line is the small custard cup, which is useful yet attractive. The salad plates may be had in the round or octagonal shape. The console set with its flare candleholders is striking and handsome. Also included in this line of black glassware are handled cake and sandwich plates, mayonnaise bowls, relish dishes and candy jars."

McKee became the McKee division of Thatcher Glass Company in 1951. In 1961 Jeanette Glass acquired the McKee Division and the McKee name disappeared completely as the site became a part of Jeannette Glass.

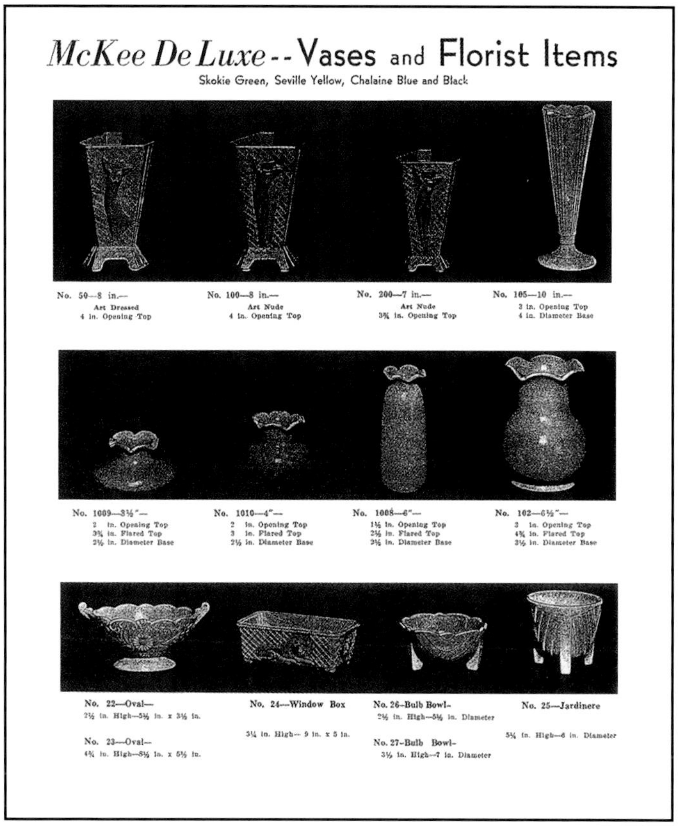

McKee Glass advertisement appearing in trade journal, 1930s.

McKee Glass advertisement appearing in trade journal, 1930.

McKee Glass advertisement appearing in trade journal, 1931.

Morgantown Glass Works

Morgantown, West Virginia
1899-1937; 1939-1971

The factory commonly known as Morgantown Glass opened in Morgantown, West Virginia, in 1899. The city was to host several dozen glasshouses over time but this one alone bears the town name. It operated continuously until 1937 during which time it underwent several changes in name and in the type of glass produced. It was first known as Morgantown Glass Works and made blown tumblers and pressed ware. In 1903 pressed ware was discontinued and the name was changed to Economy Tumbler Company. The early 1920s brought the use of color and expansion into other lines of glass so the name was changed to Economy Glass Works to better reflect a diversity of production. The company continued to operate as Economy Glass Works until its closure in 1937.

As early as 1923 *China, Glass & Lamps* reported black glass production at Economy Tumbler Co.: "...new offerings are up to the high standard of this factory...The 'Nubian' has a mirror black body with a Pomona green trimming." Again *China, Glass & Lamps* noted black glass at Economy when in the May 7, 1923, issue it stated: "The Economy Tumbler Co...other refreshment sets come in soft tints of amber, Pomona Green and black in several treatments."

Morgantown production was noted in the *Crockery and Glass Journal* February 1930, as follows: "Crystal produced by the Morgantown Glass Works, Morgantown, W. Va., appeared at the recent Pittsburgh show in many enchanting shapes and decorations. A special feature of their new stemware being the cased stem with color or black at its core (on which patent has been applied for)."

The factory reopened in 1939 as the Morgantown Glassware Guild. A few of the lines from earlier years were retained, but the focus of manufacture was institutional glass and, in later years, colored glassware of a simpler modern style. The company ceased production in 1971.

During the peak of colored production in the 1920s and 1930s, Morgantown produced baskets and vases with applied handles and feet, two-colored ware, and stemware with a fine color thread or filament inside the stem. Filament stems were, at the time, unique to Morgantown.

Black or "ebony" colored glass was made at Morgantown during the 1920s and '30s and used for a wide variety of forms and items. It was at times the body of an object and at times used in combination with crystal or colored glass. Black was made again during the Guild years and into the 1970s, but not in great supply.

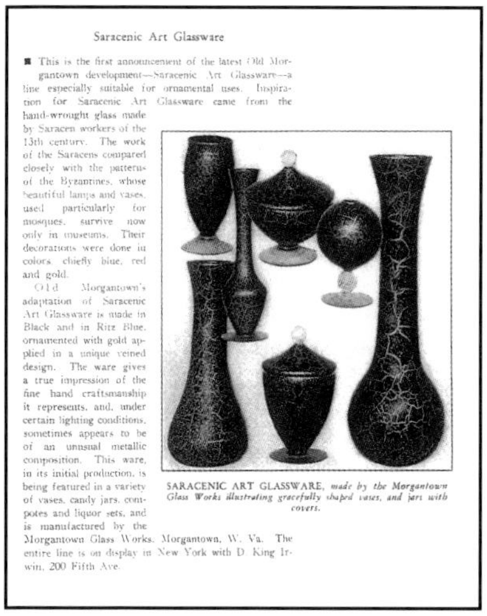

Saracenic Art Glassware

This is the first announcement of the latest Old Morgantown development—Saracenic Art Glassware—a line especially suitable for ornamental uses. Inspiration for Saracenic Art Glassware came from the hand-wrought glass made by Saracen workers of the 13th century. The work of the Saracens compared closely with the patterns of the Byzantines, whose beautiful lamps and vases, used particularly for mosques, survive now only in museums. Their decorations were done in colors, chiefly blue, red and gold.

Old Morgantown's adaptation of Saracenic Art Glassware is made in Black and in Ritz Blue, ornamented with gold applied in a unique veined design. The ware gives a true impression of the fine hand craftsmanship it represents, and, under certain lighting conditions, sometimes appears to be of an unusual metallic composition. This ware, in its initial production, is being featured in a variety of vases, candy jars, compotes and liquor sets, and is manufactured by the Morgantown Glass Works, Morgantown, W. Va. The entire line is on display in New York with D. King Irwin, 200 Fifth Ave.

SARACENIC ART GLASSWARE, *made by the Morgantown Glass Works illustrating gracefully shaped vases, and jars with covers.*

Morgantown Glass trade announcement appearing in *Crockery and Glass Journal*, October 1931.

Morgantown Glass original catalog illustration 1931 catalog. Features black cased stem lines.

Morgantown Glass advertisement appearing in *Crockery and Glass Journal*, June 1931.

Rich Colors
combined with Crystals

The Stemware and miscellaneous Vases and Basket illustrated are produced in several of the warm rich colors for which Old Morgantown is noteworthy. The 7660 Stemware is furnished in Crystal and in Ruby or Ritz Blue combined with Crystal. The 20-4" Basket, 7621-8" Flip Vase, 35½-10" Vase, with handles, and the 35½ Crimped Vase are to be had in Ruby, Ritz Blue, Ebony or the wonderful Stiegel Green with Crystal trimming.

Morgantown Glass original catalog illustration 1931 catalog.

Morgantown Glass original catalog illustration 1931 catalog.

Morgantown Glass original catalog illustration 1931 catalog.

Morgantown Glass original catalog illustration 1931 catalog.

Morgantown Glass original catalog illustration 1931 catalog.

Morgantown Glass original catalog illustration 1931 catalog.

Mosser Glass, Inc.

Cambridge, Ohio
1971 - operating at time of writing

Orie Mosser was plant manager at Cambridge Glass Company in Cambridge, Ohio, and his son, Thomas, worked there as well before Cambridge closed in 1954. Thomas went on to form his own company, Variety Glass, circa 1961. In 1971 Thomas founded Mosser Glass, which has grown to employ, in 2004, thirty people at its factory site in Cambridge, Ohio. Mosser has acquired moulds from companies including L. G. Wright and has had moulds made for their own designs to use in their pressed glass house.

Mosser production of black amethyst glass was largely confined to the 1980s and 1990s according to a factory spokesperson. Production would have been mostly limited to smaller figurals and animals, including miniature hens on the nest. While black is reportedly made "infrequently" at Mosser at this time (presumably for private mould work and special orders), it is not a color listed in active production in the company catalog.

New Martinsville Glass Co.

New Martinsville, West Virginia
1900 – 1944

Opening near the Ohio River in New Martinsville, West Virginia, in 1900 was New Martinsville Glass Manufacturing Co. Initial production included colorless pressed tableware, lamps, and novelties. English glassmaker and color specialist, Joseph Webb, joined New Martinsville at the end of 1901 and shortly thereafter a line of art glass called Muranese was on the market. Webb left in 1904 and production of the colorful ware continued for several years. However, New Martinsville's dominant lines were clear and colored pressed tableware and novelties, both before and after the Muranese line.

No mention of black glass appears in ads that mention the color availability from New Martinsville throughout the 1920s, but popular items from the 1930s can be found with black glass bases, stoppers, lids, etc. and some lines, such as the very Art Deco No. 33 "Modernistic" line are found today in black glass.

New Martinsville produced a variety of dresser and vanity sets with trays, covered powder boxes, perfumes with stoppers, and similar articles that include color and black combinations in the piece assortment. By the early 1940s, New Martinsville was promoting a line of "Viking" glass, designs influenced by clear, heavy Scandinavian glass. Colors were fading from the available color options before the firm changed ownership and name to become Viking Glass in 1944.

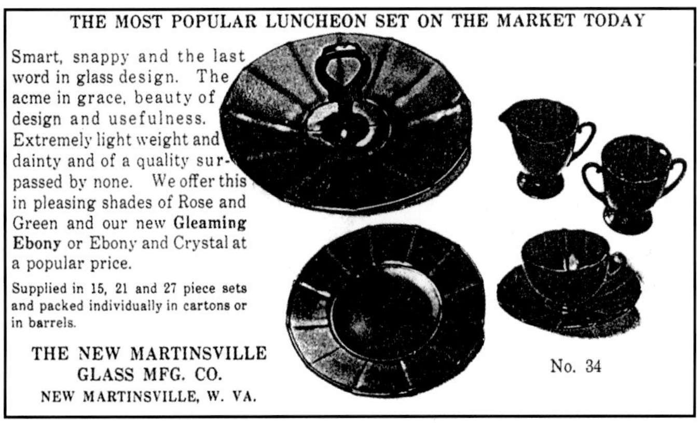

New Martinsville Glass advertisement appearing in trade journal circa 1930s.

H. Northwood Company

Wheeling, West Virginia
1902-1925

The H. Northwood Company was founded in Wheeling, West Virginia, in 1902 and was the third factory operated by the peripatetic and brilliant English glass man, Harry Northwood. His previous companies, the Northwood Glass Company in Martin's Ferry, Ohio, and Ellwood City, Pennsylvania, and the Northwood Company in Indiana, Pennsylvania, had set high standards in the production of blown and pressed glassware. His final venture was equally successful, notable for the introduction of a series of unusual opaque colors and iridescent finishes. Many of the most sought-after carnival glass patterns date to Northwood's Wheeling years. After Northwood's death in 1919, the factory continued to operate but, without his oversight, was unable to remain profitable.

Some Northwood items are marked with an uppercase letter "N", usually underlined and sometimes enclosed in a circle.

The *Crockery and Glass Journal* reported March 30, 1916, "The H. Northwood Co. have just sent Chas. J. Dela Croix, their New York representative, a new assortment of items in black glass, which make a valuable addition to the concern's already interesting display of this ware. New styles are featured in flat edge, deep flower bowls, shown in plain and pretty enamel decorations. Another article attracting more than usual attention is a combination fruit and flower holder."

Black was in the production line at Northwood as early as 1916 as *The Pottery, Glass & Brass Salesman* also reported that this "...black glass, comprising an unusually large assortment of items, is ornamented with coin gold rim treatments as well as flower patterns in enamel colors." (March 30, 1916) Many of their other colors were also available with removable black glass stands or in some cases, with applied feet of black glass.

Northwood closed in 1925.

Owens-Illinois

Toledo, Ohio
1929 - operating at time of writing

When the Owens Bottle company merged with the Illinois Bottle Company in 1929, a world class industrial giant was born. Since its formation as Owens-Illinois, the company has produced billions of bottles around the globe. Also produced have been a number of lines of commercial, industrial and utilitarian items. It would be unfair to assert that Owens-Illinois (O-I) produced significant tableware or black glass, but some information confirms that some production was undertaken. The ability to make bottles cost effectively enough by machine that they were disposable was easily adapted to things like the illustrated salt shakers which were, after all, really just small bottles with specialized lids! The extent of other O-I black glass production is not known.

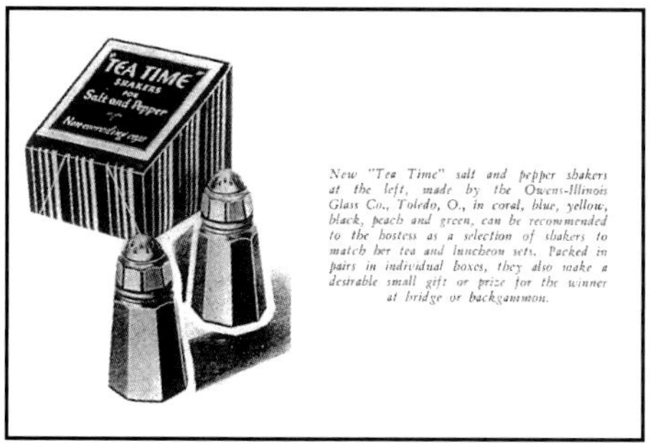

Owens-Illinois trade announcement appearing in *Crockery and Glass Journal*, September 1930.

Paden City Glass Manufacturing Company

Paden City, West Virginia
1916-1951

When David Fisher, former manager of the New Martinsville Glass Manufacturing Company, left that concern in 1916, he was offered the position of president and general manager in the newly formed Paden City Glass Manufacturing Company in nearby Paden City, West Virginia. Initially Paden City produced largely inexpensive crystal glassware for hotels, restaurants, and home use. However, by the 1920s, the company had embarked on the production of colored glassware of higher quality, offering many different styles of decoration, including some acid etchings, cuttings, and hand-painted items.

Some of Paden City's early production is indistinguishable from products also made by New Martinsville Glass Manufacturing Company. Their glassware from the 1920s to the 1940s is more easily distinguishable and identifiable. It continues to grow in popularity with collectors. Unfortunately the 1940s proved to be a difficult decade for the Paden City company. Their purchase of the American Glass Company (also located in Paden City) was an effort to compete with the machine-made glass that had flooded the market at that time, but proved too much of a financial strain and Paden City was forced to close its doors in 1951. The Canton Glass Company reportedly purchased many of Paden City's molds and all unsold inventory.

Among the many colors made by Paden City, black, also referred to as ebony at times, played a significant role. It was introduced in 1923 and continued to be produced into the 1930s.

"In black glass the company has brought out a short line which, however, includes some interesting shapes. There is a black sherbet, some odd shapes in cigar trays, and some enticing sugar and creams and candlesticks." *China, Glass and Lamps* January 22, 1923.

Some of Paden City's pieces were available with the No. 517 Black Forest etching, which was offered plain, with burnished gold bands, or with gold encrustation. Frank L. Van Deman & Son, a New York wholesaler advertised this pattern; and some sources believe it may have been produced exclusively for them in ebony. However, a report in the November 13, 1930, issue of *The Pottery, Glass & Brass Salesman* may refer to this etching: "Another new item from the Paden City Company is a new flat-top 10-inch vase made to retail at $1. The vase is in black, and on the flattened flange, in a position of great visibility, is a hunting scene etching of modernistic motif."

The same trade journal also reports that: "One of the smartest of the new items is a three piece batter set that, in shape and color, is as trim and eye-arresting as the stream-line hull of a racing sloop. The set is composed of a dough-and-syrup pitcher and tray. The pitchers, of svelte, graceful mold, are in crystal. The tray and pitcher tops are in black. The result is a vivacious, thoroughly modern ensemble of utility and beauty that guarantees consumer approval."

The Pairpoint Corporation

New Bedford, Massachusetts
1894-1938

Two companies in New Bedford, Massachusetts, merged to form one concern July 14, 1894, when Pairpoint Manufacturing Company (founded 1890) acquired the assets of the Mt. Washington Glass Company (founded 1871). Both had had overlapping ownership and had worked for many years to join the silver-plated metal works of Pairpoint with the glass of Mt. Washington. The companies were working in tandem when reorganized in 1900 to form the Pairpoint Corporation. Pairpoint closed in 1938 to re-open in 1939 as Gunderson Glass Works. Gunderson remained in production until 1952.

Black was produced at Pairpoint but is relatively uncommon compared to their other colors. Black was frequently combined with crystal where the stems, feet, and bowls were black. Black was less commonly combined with other colors, like their vibrant red, called Flambeau. Pairpoint pieces with a controlled bubble ball are some of their more recognizable work. A note of caution is needed to avoid confusion with similar products with air trap bubble stems and bases by Erickson and Tiffin glass from Ohio as well as other hand glass plants.

Pilgrim Glass Corporation

Ceredo, West Virginia
1948-2001

Alfred Knobler, once a sales representative selling handmade glass from West Virginia, acquired the factory of Tri-State Glass and created his own Pilgrim Glass in Ceredo, West Virginia, in 1948. He retained control of the factory until its closing in April 2001. Pilgrim was one of several small to medium sized handmade glass plants in the Teays Valley region of West Virginia creating similar and often confusingly similar products. For a long time the dominant products were small and brightly colored pitchers, vases, and such for window ledges in homes.

Pilgrim went on post 1970s to distinguish itself with products such as a line of gigantic proportions called the "Masterworks" in colorful hand blown forms and a world-class line of hand carved Cameo glass under the direction of art director Kelsey Murphy.

Black glass was mastered and entered the line at Pilgrim in 1987. It was included in the production of mainline catalog items, the Masterworks line of oversized pieces, and in the skillfully hand-formed layering of some Pilgrim Cameo glass. The factory closed in 2001.

Note that when Pilgrim closed, the West Virginia Museum of American Glass purchased the remaining stock of original company glass catalogs and these original catalogs are, and will be as long as supplies last, available for purchase from the museum. See contact information in the introductory chapter of this book.

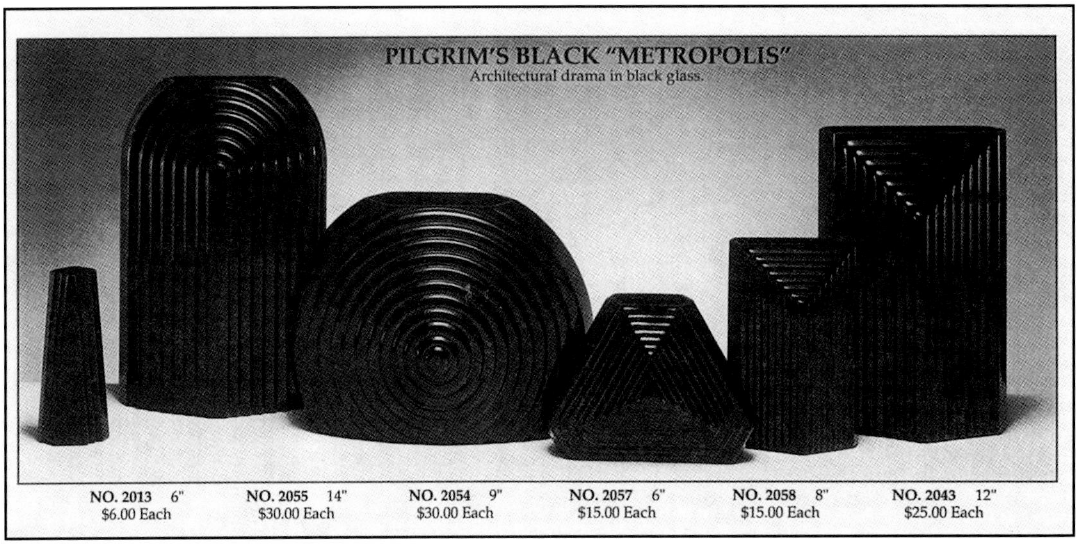

Pilgrim Glass original 1994 catalog illustration.

Pilgrim Glass original 1994 catalog illustration. These hand-painted items appear in the 1994 catalog only.

Pilgrim Glass original 1994 catalog illustration. Note the columns shown were also made in a third size, between the two shown here.

Pilgrim Glass original 1996 catalog illustration.

Pilgrim Glass original 1994 catalog illustration. The same images appeared in 1995 but not 1996.

Pilgrim Glass original 2000 catalog illustration.

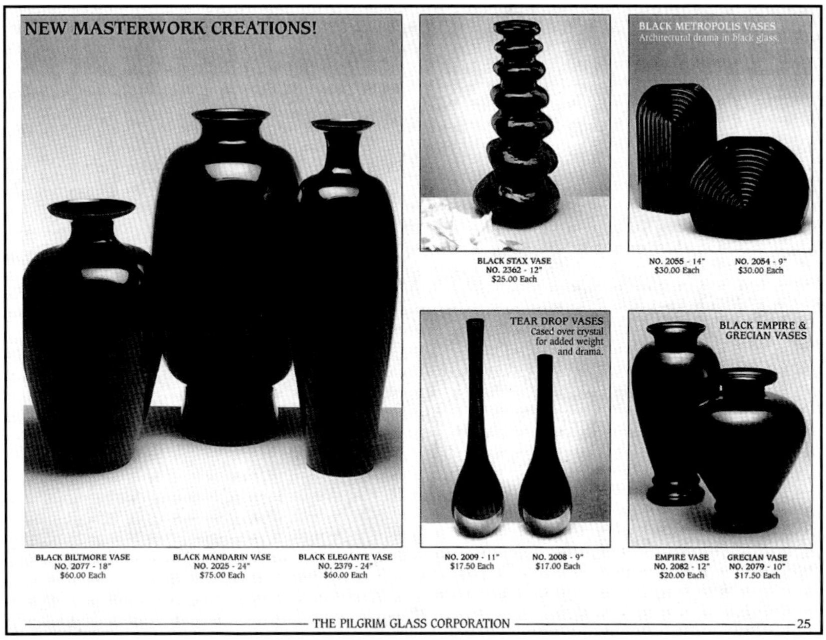

Pilgrim Glass original 2000 catalog illustration.

Pilgrim Glass original 2000 catalog illustration.

Plum Glass Company

Pittsburgh, Pennsylvania
1984 - circa 2002

The Plum Glass Company was established in Pittsburgh, Pennsylvania, in 1984 by John J. Biondich. He had been an employee of the Westmoreland Glass Company for twenty-eight years at the time of Westmoreland's closing in the 1980s and was able to purchase a number of Westmoreland molds. He used these molds, returning the pieces to active production.

With limited production capacity, Plum Glass gave new life in a variety of colors, including black, to familiar Westmoreland shapes and patterns. This was a small family operation with John, his brother, George, and three sisters, Helen, Annie, and Zora, handling everything from mixing the batch to hand pressing the glassware.

John Biondich died in 2002. Glass Directories as late as 2004 continue to list the company.

Scandia Glass Works, Inc.

Ceredo, West Virginia
1946-circa 1983

Established in 1946 and operating for several decades, this medium sized glass company made colored, handmade glass. It is little known or completely unknown in today's collecting market. In 1976 Scandia Glass employed forty-four workers and remained in the control of the Bergquist family who founded it.

The West Virginia Museum of American Glass was given a large selection of objects that remained in the

Scandia factory at closing. This important collection includes crackle small vases and pitchers, hand formed animals, and other objects. Shown in the animal chapter of this book are a few of the glass animals from Scandia that were made of black glass. Whales, birds, cats, mice, and more are known in the Scandia animal line.

Scandia closed circa 1983.

Seneca Glass Company

Morgantown, West Virginia
1896-1982

Seneca Glass was the first glass house to open in Morgantown, West Virginia, when it lit its furnaces there in 1896. Many others would follow. Seneca had begun in 1891 in Seneca County, Ohio, from which the glass company takes its name, but relocated when Ohio fuel supplies presented challenges. The founders and owners of the company were largely German immigrants from the same region in the Black Forest part of Germany. Some of those same families were in leadership roles almost a century later near the end of the company's long life.

Seneca began its production making high end lead crystal, much of it cut in the heavy Rich Cut or Brilliant Cut style of the later Victorian era. By the nineteen teens a lighter, thinner, more spartanly cut line had also entered the Seneca offerings. Elaborately cut and expensive glass would remain in the catalog lines until the last decade of production. Black glass appears in the 1920s at Seneca as trim on colorless vessels – feet, handles, stems of goblets, etc.

The 1950s saw the beginning of the end as Seneca responded to popular demand and introduced casual glassware. The days of elegantly cut, expensive fine crystal were numbered. In 1982 Seneca was sold by the heirs and successors to the founders and purchased by a group of investors from Malaysia. Then named Seneca Crystal Incorporated, the firm lasted less than two years. Assets, including a near century's worth of samples and equipment, were sold at public auction. The factory building survives today as a well preserved specialty shopping mall with the worn floors, aged brick, and other relics of the once busy factory still to be found in the structure if you look closely enough.

Sloan Glass

Lonaconing, Maryland
1930-1932

The Sloan brothers, Dixon and Alexander, relocated their efforts to a defunct glass factory in Lonaconing, Maryland, in 1931 after their Potomac Glass Co. burned down. While briefly at the Lonaconing site, they continued some of their products from the Potomac Glass Co. as well as adding some new products. Among these is a three sided concave stem patented during the Lonaconing time shown in the stemware chapter.

The Lonaconing factory of Sloan Glass burned in 1932.

L.E. Smith Glass Company

Mt. Pleasant, Pennsylvania
1907-operating at the time of this writing

The early history of the L.E. Smith Glass Company, as frequently recounted, is colorful, if slightly at odds with the historical record. For example, Smith recalled their initial years in a catalog issued in the 1990s as follows, "Starting long ago, in 1907, Lewis Smith had a hobby of making his own mustard. When confronted with the problem of what to put his mustard in, the first L. E. Smith glass tank was built. From the seeds of mustard making, Lewis Smith harvested what has become a nearly century of rich traditions and excellent craftsmanship." The mustard part of the story is quite true, but it ignores the fact that L.E. Smith actually became involved in the glass business in the early 1890s, serving, among other capacities, as manager of the decorating department of the McKee-Jeannette Glass Company in nearby Jeannette, Pennsylvania. It was those facilities that were leased in 1907 when Smith struck out on his own as a decorator. In 1909 he purchased the factory of the bankrupt Anchor Glass Company in Mt. Pleasant, Pennsylvania, and began glass production. This included containers for "Smith's German Mustard."

An extensive variety of utilitarian, tableware, and decorated pieces were made by Smith over time. In the 1920s they expanded their production with the purchase of the Greensburg Glass Works and its molds, with both factories operating for a number of years under Smith's control. In the 1940s they began emphasizing reproductions of antique glassware, including the patterns Moon and Star, Daisy and Button, and others. In the 1960s, when glass sales and collecting were booming, some of their products were distributed under the name of Fayette Glass, an extension of their usual distribution channels. In 1975, Owens-Illinois bought the Smith factory, continuing to operate it as an independent subsidiary. When they sold the factory in 1986, it was purchased by a group of local investors who operated it until its closing in 2004. As of this writing, the factory has re-opened under new ownership.

Smith has offered many colors over the near century, with black being one that recurs frequently. It was first introduced at Smith in the late 1920s and became one of Smith's most popular selling colors by the 1930s. Pieces from this period will often be decorated, usually with silver, but also with enamel designs. Black glass was offered again in the 1950s and from this period it may be found with an unusual marble-like flecking. Black was again in the line in the early 1980s and from 1988 to 1992, when it was at times referred to as ebony.

L.E. Smith Glass advertisement appearing in trade journal circa 1930s.

L.E. Smith Glass original catalog page circa 1930s.

No. 600. Double Candlestick Holder
Black or Blue
Packed 4 doz., Wt. 55 lbs.

No. 2400. 2 Handled Footed Fruit Bowl
Black or Blue
Packed 2 doz., Wt. 65 lbs.

No. 1022. 3 Footed Console Bowl
Black or Blue
Packed 2 doz., Wt. 55 lbs.

No. 49. Crimped Top
2 Handled Footed Vase
Black with Silver Dec.
Packed 4 doz.
Wt. 45 lbs.

No. 102. Fancy
2 Handled Vase
Black with Silver Dec.
Packed 4 doz.
Wt. 60 lbs.

No. 1900. Flared Top
2 Handled Footed Vase
Black or Blue
Packed 2 doz.
Wt. 50 lbs.

No. 432/5. Fancy
Crimped Top
2 Handled Footed Vase
Black or Blue
Packed 2 doz.
Wt. 33 lbs.

No. 433. Fancy
Crimped Top
2 Handled Footed Vase
Black or Blue
Packed 2 doz.
Wt. 40 lbs.

No. 85. Hobnail
Footed Ivy Ball

No. 1. 3 Footed
Fern Bowl

No. 2400. 2 Handled
Footed Bon Bon Dish

No. 1000
Cockerel
Ash Tray

No. 1000
Scotty Dog
Ash Tray

No. 66. Flower Pot
and Saucer
Black with Silver Dec.

L.E. Smith Glass original catalog page circa 1930s.

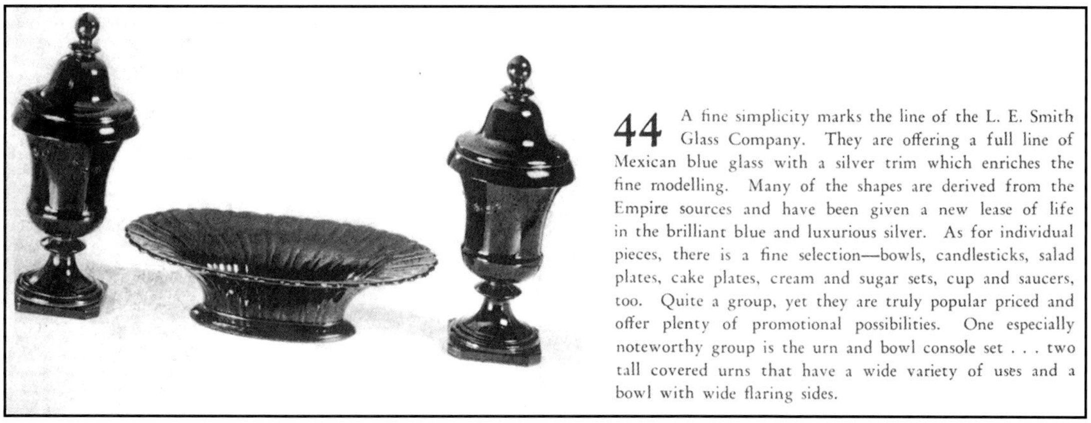

44 A fine simplicity marks the line of the L. E. Smith Glass Company. They are offering a full line of Mexican blue glass with a silver trim which enriches the fine modelling. Many of the shapes are derived from the Empire sources and have been given a new lease of life in the brilliant blue and luxurious silver. As for individual pieces, there is a fine selection—bowls, candlesticks, salad plates, cake plates, cream and sugar sets, cup and saucers, too. Quite a group, yet they are truly popular priced and offer plenty of promotional possibilities. One especially noteworthy group is the urn and bowl console set . . . two tall covered urns that have a wide variety of uses and a bowl with wide flaring sides.

L.E. Smith Glass trade journal announcement appearing in *China and Glass*, January 1935.

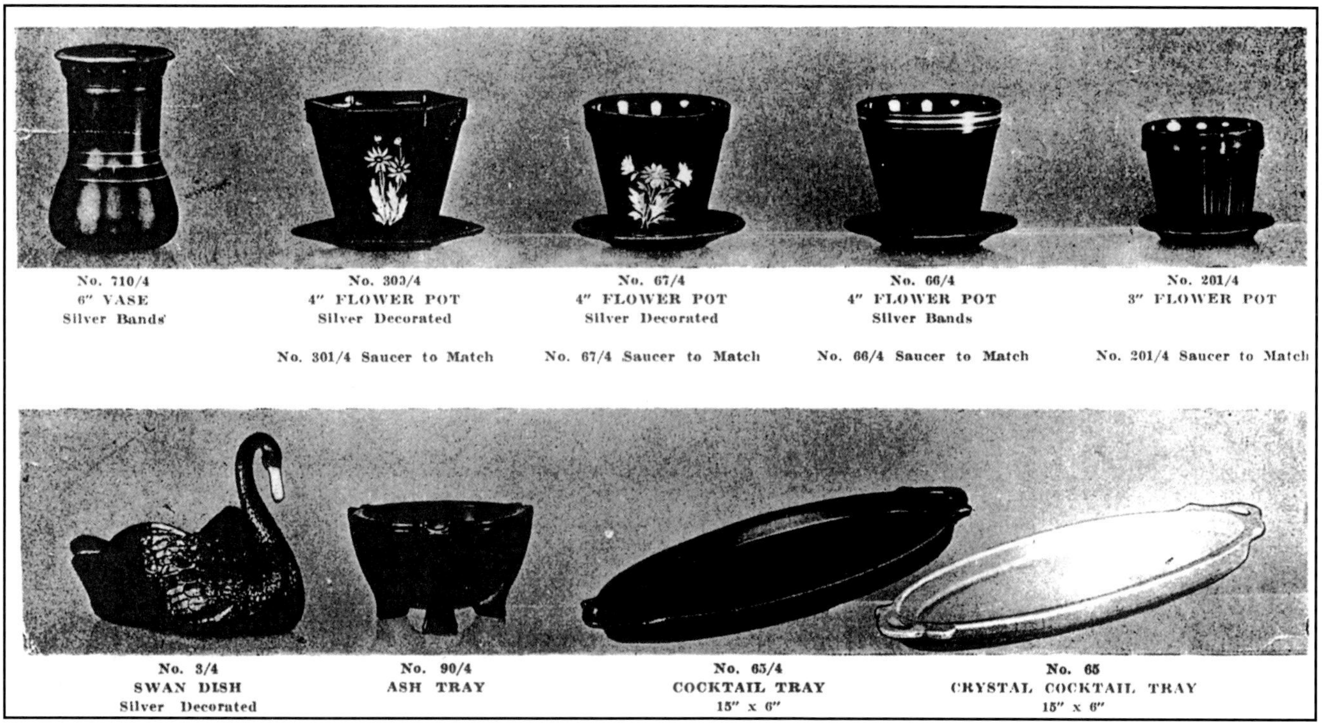

No. 710/4 6" VASE Silver Bands
No. 303/4 4" FLOWER POT Silver Decorated
No. 67/4 4" FLOWER POT Silver Decorated
No. 66/4 4" FLOWER POT Silver Bands
No. 201/4 3" FLOWER POT
No. 301/4 Saucer to Match
No. 67/4 Saucer to Match
No. 66/4 Saucer to Match
No. 201/4 Saucer to Match
No. 3/4 SWAN DISH Silver Decorated
No. 90/4 ASH TRAY
No. 65/4 COCKTAIL TRAY 15" x 6"
No. 65 CRYSTAL COCKTAIL TRAY 15" x 6"

L.E. Smith Glass original catalog page circa 1930s.

No. 505 One Handled Sandwich Tray 10" Diameter
No. 505 Two Handled Salad Bowl 8" Diameter—3½" High
No. 505 One Handled Nut 8¾" Diameter
No. 505 Two Handled Salad Tray 10½" Diameter
No. 309 Flared Console Bowl 12½" Diameter
No. 309 Cupped Bowl 8¾" Diameter—5½" High
No. 327 Cupped Bowl 6¾" Diameter—5" High
No. 404 Window Box and Block 8" Long—3¾" Wide—3¾" Hi
No. 515 Footed Nut Bowl 9" Diameter
No. 1 Fern Bowl 6¼" Diameter—5½" High
No. 515 Footed Cake 10½" Diameter
No. 515 Footed Cupped Bowl 7" Diameter—3¾" High
No. 1000 Flared Vase 11" Diameter—6" High
No. 1000 Cupped Vase 6½" Diameter—7¾" High
No. 1000 Fan Vase 10" Diameter—8" High
No. 1931 Vase 4½" Diameter—7¾" High
THE L. E. SMITH GLASS COMPANY Mount Pleasant, Pa.

Left:
L.E. Smith Glass original catalog page circa 1930s.

Right:
L.E. Smith Glass original catalog page circa 1930s. Note these are items cited as "Special Listing with F.W. Woolworth," a national variety store chain of the era.

L.E. Smith Glass trade journal advertisement 1932.

L.E. Smith Glass trade journal advertisement, circa 1930s.

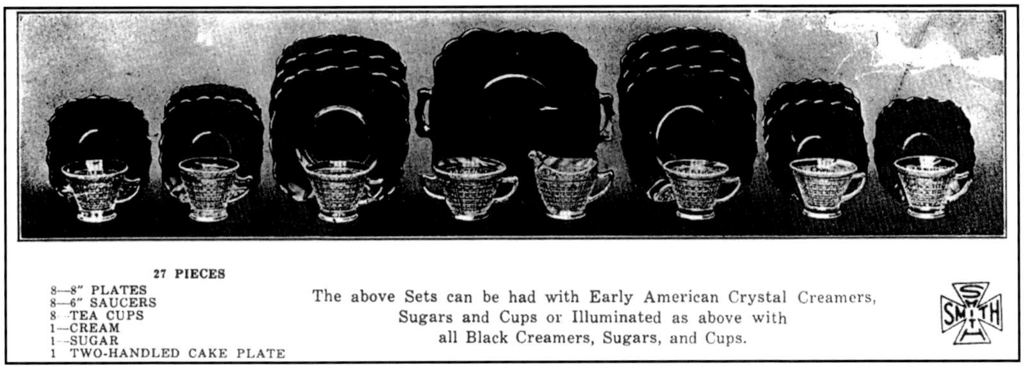

L.E. Smith Glass illustration for Early American. Circa 1930s.

Standard Glass Manufacturing Company, see Lancaster Glass Co.

Steuben Glass Works

Corning, New York
1903 - operating at time of writing

The Steuben Glass Works was begun by Frederick Carder, an English glass worker and artisan. Carder opened Steuben in 1903. By late 1904 Carder began making the first of his art glass – Aurene being the first to be made with many to follow.

While successful for a period, the company was sold to Corning Glass Works in 1918 at the end of World War I. Corning retained Carder as the managing director of the company and for years he was given almost free reign to develop colors and create artistic objects.

The period of interest to black glass enthusiasts began in the 1920s. Two examples of Steuben black are Intarsia, an art glass line of black designs picked up in the hot glass forming process, and reeded designs (or threaded) where literal threads of black are wrapped around the body of a differently colored object. Objects made include creams, sugars, cocktail sets, dresser sets, vases, and pitchers, as well as others.

Steuben is operating today, crafting very high quality (and expensive) crystal objects, often hand-engraved with elaborate designs. No black or colored glass has been produced commercially at Steuben for decades.

Sowerby's Ellison Glass Works

Gateshead-On-Tyne, England, UK
1807-1972

Origins are not clear, but it appears the glass factory was established about 1807. As with many factories, names and persons involved in the company changed throughout the following years. It was known as the New Stourbridge Glassworks from 1807-1852.

The Gateshead Stamped Glassworks produced pressed glass exclusively (one of the first in England to make pressed glass), but was closed in 1848.

The above were precursors of the Ellison Glassworks which was in business from 1852 to 1972. The Sowerby family was involved in all three factories.

Sowerby is well known for its opaque glass in Blanc de Lait (milk glass), turquoise, Queen's Ivory (custard), malachite (mosaic or slag glass), blue malachite, green malachite, tortoiseshell, and others. Jet or black glass was also made in the 1870s and later. Examples of Sowerby pressed glass generally exhibit sharp and detailed mold work and include many forms not commonly made in America as well as more commonly seen pieces.

Tarentum Glass Co.

Tarentum, Pennsylvania
1893-1918

The Tarentum Glass Company came into existence in 1893 when a group of local businessmen purchased the old Richards and Hartley factory in Tarentum from the U.S. Glass Company. As Tarentum Glass, they produced a number of tableware patterns and novelties. In 1913 the factory was closed, but reopened a year later. For the next few years production was devoted primarily to lighting goods, including candlesticks. When the factory burned in 1918, a decision was made not to rebuild.

"The Tarentum Glass Co. are constantly adding to their line of candlesticks, with the result that the assortment on view at the salesroom of their agent, W. F. Rood, offers as fine and varied a selection as can be found on the market. Aside from the number of designs and styles in plain and decorated crystal, there is a choice and diversified showing of dainty and novel designs in opal glass. Included among those are several odd black and white patterns that are good; then there are blue, pink, and other delicately colored designs that make a pretty contrast to the opal glass." *Crockery & Glass Journal,* March 30, 1916.

Tiara, see Indiana Glass

Tiffin, see U.S. Glass below

U.S. Glass Company

Pittsburgh, Pennsylvania
1891-1984

The massive U.S. Glass Company combine was organized in 1891 by the corporate joining of several existing glass companies:

Factory A – Adams & Co., Pittsburgh, PA.
Factory B – Bryce Bros., Pittsburgh, PA.
Factory C – Challinor, Taylor & Co., Tarentum, PA.
Factory D – Geo. Duncan & Sons, Pittsburgh, PA.
Factory E – Richards & Hartley, Tarentum, PA.
Factory F – Ripley & Co., Pittsburgh, PA.
Factory G – Gillinder & Sons, Greensburg, PA.
Factory H – Hobbs Glass Co., Wheeling, WV.
Factory J – Columbia Glass, Findlay, OH.
Factory K – King Glass Co., Pittsburgh, PA.
Factory L – O'Hara Glass Co., Pittsburgh, PA.
Factory M - Bellaire Goblet Co., Findlay, OH.
Factory N – Nickel Plate Glass Co., Fostoria, OH.
Factory O – Central Glass Co., Wheeling, WV.
Factory P – Doyle & Co., Pittsburgh, PA.
Factory R – A.J. Beatty & Sons, Tiffin, OH.
Factory S – A.J. Beatty & Sons, Steubenville, OH (inoperative by time of combine)

With so many companies involved in the merger that formed U.S. Glass, almost any type of glass was available from U. S. Glass, but the product line was primarily tableware, druggist wares, lamps, and barware at the time of the formation. In 1938 company offices were moved from Pittsburgh, PA, to Tiffin, OH, however the name U.S. Glass was maintained.

Black satin glass was a huge success for U.S. Glass during the 1920s with most of it being made at Factory K (King) in Pittsburgh. Black satin was advertised as early as 1923. U.S. Glass sold so much black satin glass that at the time of this writing, decades later, it is a safe beginning point when identifying a piece of black satin glass to start by looking at U.S. Glass as the possible manufacturer. Black glass was made in console sets, novelty items, vases, smoker's items, and a multitude of other pieces.

Towards the end of the production at U.S. Glass, several company names were used and various owners attempted to keep it open. Operating as Tiffin Art Glass Corporation, a few lines of all black stemware were produced. The Tiffin plant was last of all the U.S. Glass Company sites to operate when it closed in 1984.

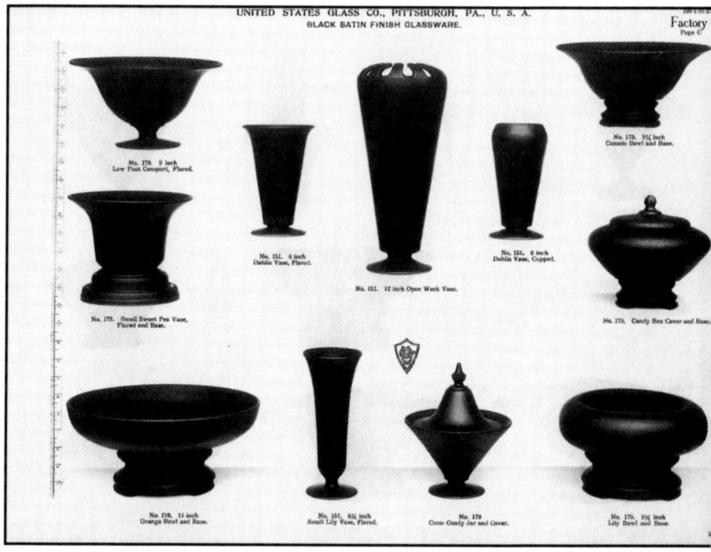

U.S. Glass original catalog page. Dated January 1927. *Compliments Tiffin Collectors Club.*

U.S. Glass original catalog page. Circa 1920s. Products from Factory "D." *Compliments Tiffin Collectors Club.*

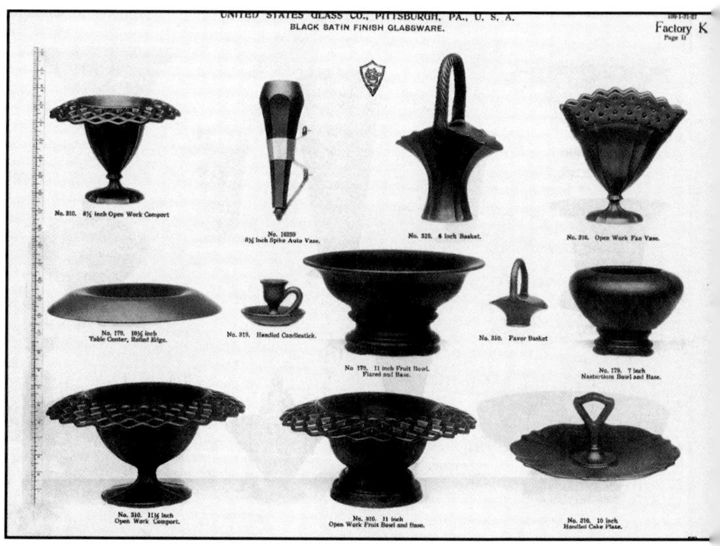

U.S. Glass original catalog page. Dated January 1927. *Compliments Tiffin Collectors Club.*

Viking Glass Company

New Martinsville, West Virginia
1944-1986

Viking Glass was formed when the New Martinsville Glass Co. of New Martinsville, West Virginia, closed. Viking began in 1944 and chose its name to emphasize the colorless and heavy styled crystal it was then producing. This style of crystal was in the manner of Scandinavian glass, thus the word Viking. The glass at the time was popular for its modernity.

Viking produced black or "ebony" glass from 1949 until 1956. A black glass with a dull finish (satin glass?) was promoted in a "What's new for the July market" article in *Crockery & Glass Journal*: "Viking Glass Co., New Martinsville, W.Va. Everybody ...was excited about a new black glass with a dull finish which is to be called MIDNITE. Included in the new line are smoker items, bowls, candlesticks, vases and many other items to retail for 50c to about 3$." July 1954.

Black was reintroduced after 1980 when the factory was operating as Dalzell-Viking. Viking forms were made by Dalzell-Viking in the same molds that remained in the factory from earlier Viking years, making distinguishing the glass, at times, almost impossible.

Viking Glass original catalog page 1977-78 catalog.

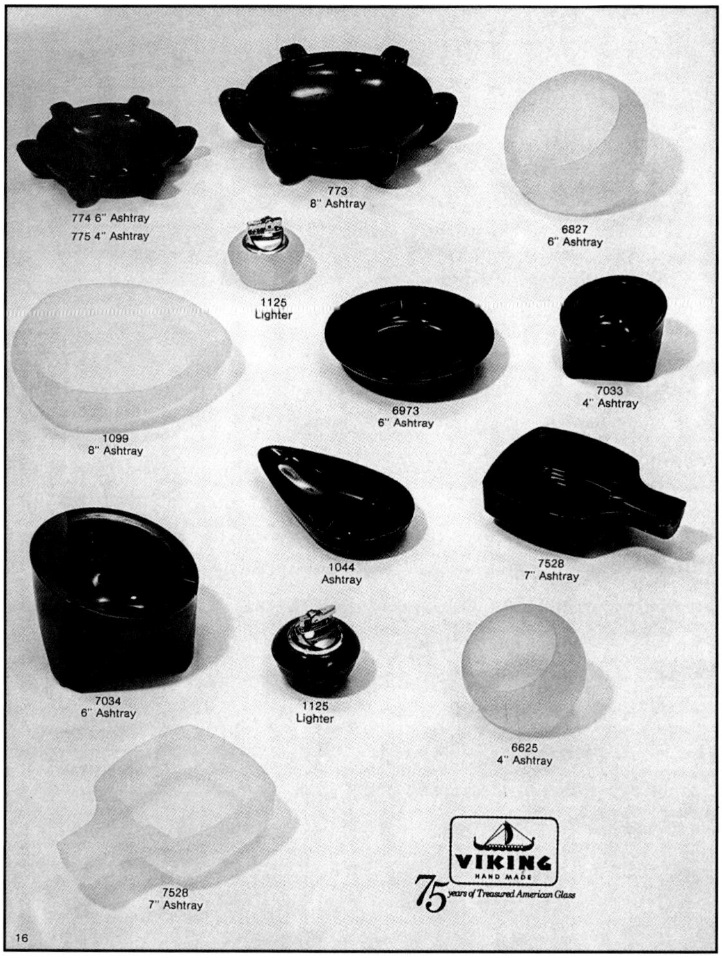

Viking Glass original catalog page 1975.

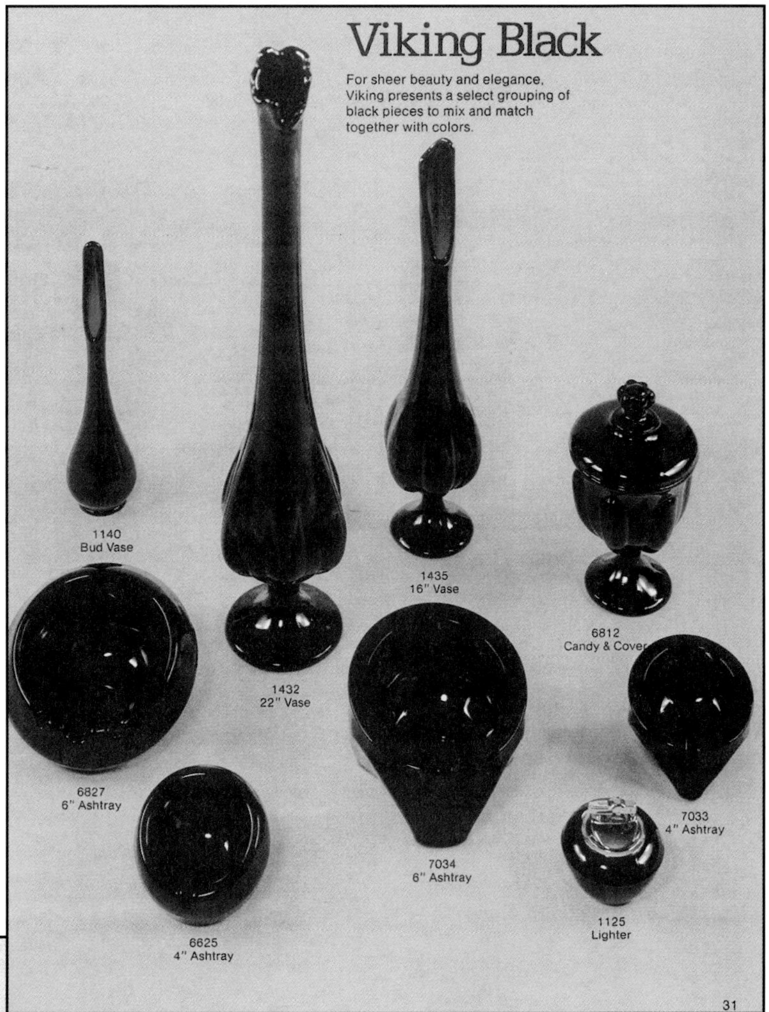

Viking Glass original catalog page 1979-80 catalog.

Viking Glass original catalog page 1975.

The Vitrolite Company

Parkersburg, West Virginia
1910-1935

Founded circa 1909 as Carter-Meyercord Company in Parkersburg, West Virginia, to produce flat slabs of glass for Chicago sign maker George R. Meyercord, the firm became The Vitrolite Company in 1910.

The slabs, from 5/10th of an inch to 1 1/4 inch thick, were utilized for signs, architectural siding, and numerous projects from furniture to paperweights. A common use was to add a basalt black or colored front to Art Deco stores and in kitchens and baths to give the impression of marble walls, floors, counters, and such. Colors produced at Vitrolite were many, from black to agates, blues, greens including jade, red, peach, and white.

One impressive glass trivet or tile from Vitrolite, with the design cut or sand carved into the surface, has an art deco female nude curled, ball-like, to cover the circular tile. Uncommon pieces offer collectors challenges due to the variety. Other firms made similar glass slabs for architectural purposes but none are reportedly used for the variety of things Vitrolite was.

Libbey-Owens-Ford acquired the Vitrolite Company in 1935 and continued production of the product, still called Vitrolite, into at least the 1950s.

Westite Company

Weston, West Virginia
1930-1936

Operating as Balmer-Westite, circa 1929-30, the firm becomes Westite Company in June of 1930. Balmer was an industrialist and capitalist from New Jersey whose company had gained success in marketing porcelain bathroom fixtures. Glass bathroom fixtures were also a Balmer line. Fixtures held toothbrushes, tumblers, toilet tissue rolls, and bath towels using pressed glass bars and fixtures that were made to compliment clay or glass tile. The Weston site manufactured the glass Balmer bathroom components and later other novelties and small objects. Transparent glass was produced but the majority of the glass was opaque. The Westite color selection included black glass.

After fire destroyed the factory in 1936 production was continued, in many of the same molds, at the Akro Agate factory in nearby Clarksburg, West Virginia. Some original Westite lines, such as glass flowerpots, proved to be long lived at Akro.

Westite bathroom fixtures as illustrated in a Williams Hardware Co. catalog circa 1920s. image found in *The Complete Line of Akro Agate*. (Hardy 1992). *Used with permission of Roger and Claudia Hardy.*

The Williams Hardware Co. Clarksburg, W.Va. 89

BATHROOM FIXTURES, SCALES AND TOILET PAPER

WESTITE Bathroom Fixtures

Made of a highly fired composition as durable and sanitary as marble, the color being a part of the composition. Fitted with a concealed attachment. Furnished in white, black, jade green and baby blue. Specify color desired.

Westite Assortments

Consists of one each of the items listed below. A plush covered display board will be furnished with every order for four assortments.
Westite Assortment..........Each $30.00

Towel Bars

Length, in. 18 24
No. W90—Each $5.00 $5.50

Towel Bars With Center Post

No. W92—Length 36 in......Each $8.50
No. W92—Length 48 in......Each 9.50

Tumbler Holders

No. W84—Tumbler Holder....Each $1.50

Tumblers

No. W83—Tumbler ....Each $0.75

Soap Dishes

No. W81—Soap Dish.........Each $2.00

Toothbrush and Paste Holders

No. W80—Toothbrush and Paste Holder. ....Each $2.00

Garment Hooks

No. W87—Garment Hook.....Each $1.25

Toilet Paper Holders

No. W86—With Westite Roller..Each $3.50
No. W86—With Nickel plated spring roller...................Each $4.00

Westite original promotional photographs circa 1930. *Property West Virginia Museum of American Glass. Gift to WVMAG from the Robert Henderson family in memory of John Henderson.*

Westite original promotional photographs circa 1930. *Property West Virginia Museum of American Glass. Gift to WVMAG from the Robert Henderson family in memory of John Henderson.*

Westite original promotional photographs circa 1930. *Property West Virginia Museum of American Glass. Gift to WVMAG from the Robert Henderson family in memory of John Henderson.*

Westmoreland Specialty Company

Grapeville, Pennsylvania
1889-1984

Charles H. and George R. West, with George M. Irwin, founded the Westmoreland Specialty Company in 1889 when they purchased the abandoned Liverpool Glass Works in Grapeville, Pennsylvania. Among other products the early wares included jars for the mustard and vinegar produced by the West Brothers Grapeville Vinegar Works, located on the factory grounds. By the mid-teens they had expanded to include many complete table lines and offered cutting and hand-painted decorations. This trend was continued into the 1920s when color became dominant. In 1925 the factory was renamed the Westmoreland Glass Company honoring the county in which it was located.

Black glass was in production as early as 1916 when *Crockery & Glass Journal* reported in the issue of February 24 that: "One of the most noteworthy of the recent achievements of The Westmoreland Specialty Co. is their new line of flower pans, bowls, standards and other specialties, which bring out some exceedingly original ideas in shapes. They are made of black, Belgian blue and white carrara marble glass, with distinctive decorations. Some of the treatments are unusually beautiful, several raised gold designs on both the blue and black glass being very striking. Other decorations are raised white enamel and red enamel patterns; also a variety of hand-painted designs showing off some floral treatments. There are besides a variety of deep etchings that are rich and extremely effective."

The 1930s were difficult for Westmoreland as for most of the glass companies and the country. Under the leadership of James J. Brainard, who assumed control of the company in 1937, the new emphasis was changed to "Authentic Handmade Reproductions," with many "new" patterns introduced that had been copied from nineteenth century designs. Throughout this period and well into the 1950s milk glass was the mainstay of Westmoreland. At its peak of popularity in 1957, 95% of Westmoreland's entire output was milk glass. Transparent colors were reintroduced in the 1960s, but sales were only modest over the next two decades. In 1980 the company was sold to David Grossman, an importer and distributor from St. Louis, who continued to operate the plant until January 8, 1984 when production was halted for the final time.

In addition to Westmoreland's tremendously popular milk glass and other opaque colors black also appeared frequently in their catalogs, initially from 1915 until 1937 (when it was at times called Swedish black), and then again off and on from 1949 to 1975. A report in the November 25, 1914, issue of *The Pottery, Glass & Brass Salesman* described the then new black glass as: "Probably the most noteworthy item in the line is a matt black of a richness almost inconceivable in pressed glass...Among the decorations are gold encrustations and flowers in a variety of colors. The plain matt black and black with gold encrustations are unquestionably the most beautiful and give to the trade something absolutely unique in goods of this character."

The range offered was emphasized in another report on December 30, 1915, from the same journal: "Among the new things in black glass are several styles of footed compotes besides bud vases, handled and plain candlesticks, finger bowls and plates, ash receivers and coasters. Several of these are shown in a matt black finish as well as in the regular bright finish, and in addition there are many decorative treatments. Indeed, the showing of decorated black glassware constitutes an array that is impressive, both for variety of patterns and for the uniqueness of effects. Among these decorations are some handsome floral effects in coin gold, which are applied to a variety of pieces, including the flower bowls, banquet pieces, etc. Then there are some distinctive designs in white enamel and in white and red enamel, providing the buyer an extensive range of styles from which to make his selection."

In the 1920s crystal was offered spray-cased with black, a technique involving coating the black with finely ground glass "dust" and then reheating the piece to permanently bond the coating. Some black glass was also offered in the 1930 with a moire satin finish. An advertisement in the *Gift & Art Buyer* of March 1949 announced Westmoreland black glass offerings of turtles, stemware, and more, declaring black is back "after a long absence." In 1916 a rainbow carnival finish was offered on black, and again in 1984 some black was made for the Levay Distributing Company with a carnival or iridescent finish.

As late as 1967 Westmoreland was able to advertise on the cover of the February issue of *China, Glass & Tableware* "Black is back – and with more glamour and solid appeal than ever before ... in response to a surprising spontaneous demand."

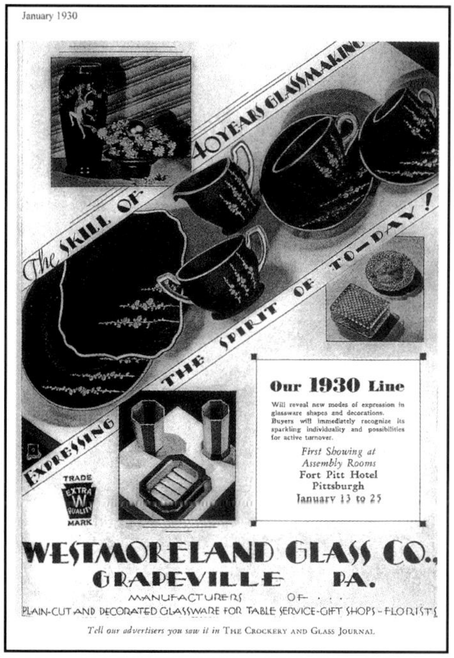

Westmoreland Glass original advertisement appearing in *Crockery and Glass Journal*, January 1930.

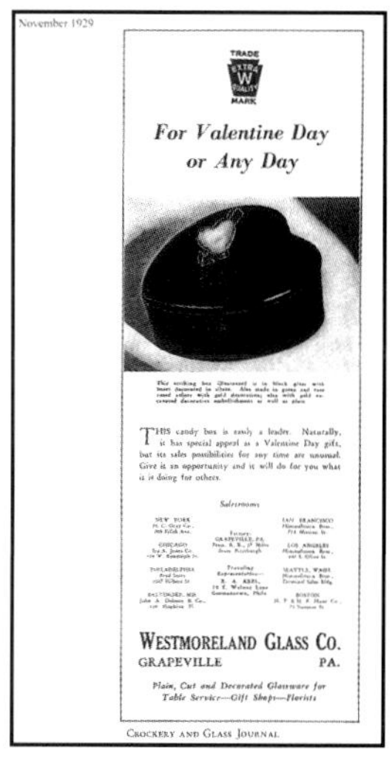

Westmoreland Glass original advertisement appearing in *Crockery and Glass Journal*, November 1929.

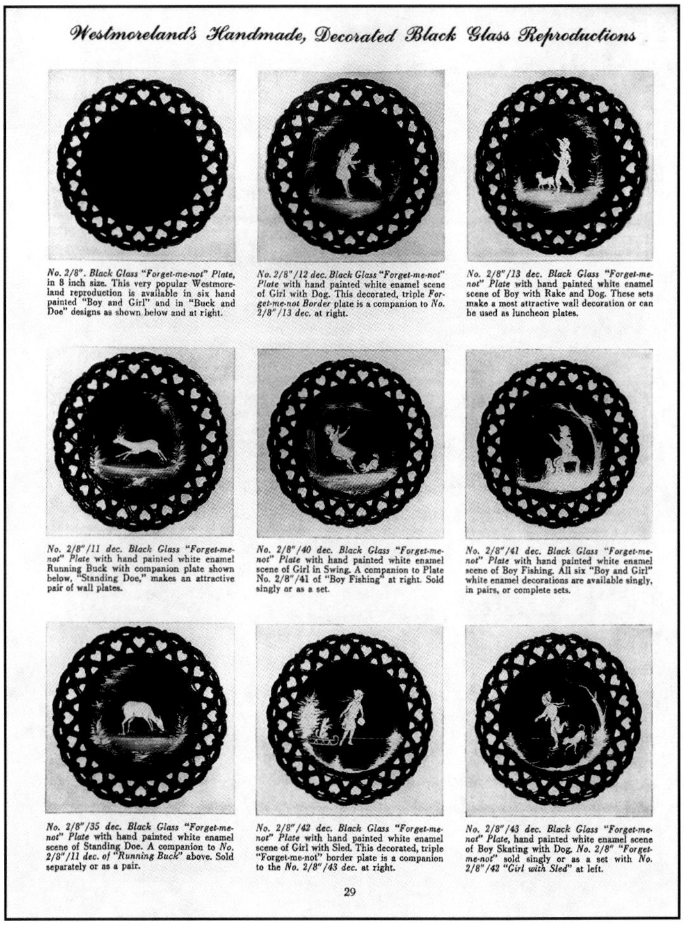

Westmoreland's Handmade, Decorated Black Glass Reproductions

No. 2/8". Black Glass "Forget-me-not" Plate, in 8 inch size. This very popular Westmoreland reproduction is available in six hand painted "Boy and Girl" and in "Buck and Doe" designs as shown below and at right.

No. 2/8"/12 dec. Black Glass "Forget-me-not" Plate with hand painted white enamel scene of Girl with Dog. This decorated, triple *Forget-me-not Border* plate is a companion to *No. 2/8"/13 dec.* at right.

No. 2/8"/13 dec. Black Glass "Forget-me-not" Plate with hand painted white enamel scene of Boy with Rake and Dog. These sets make a most attractive wall decoration or can be used as luncheon plates.

No. 2/8"/11 dec. Black Glass "Forget-me-not" Plate with hand painted white enamel Running Buck with companion plate shown below, "Standing Doe," makes an attractive pair of wall plates.

No. 2/8"/40 dec. Black Glass "Forget-me-not" Plate with hand painted white enamel scene of Girl in Swing. A companion to Plate No. 2/8"/41 of "Boy Fishing" at right. Sold singly or as a set.

No. 2/8"/41 dec. Black Glass "Forget-me-not" Plate with hand painted white enamel scene of Boy Fishing. All six "Boy and Girl" white enamel decorations are available singly, in pairs, or complete sets.

No. 2/8"/35 dec. Black Glass "Forget-me-not" Plate with hand painted white enamel scene of Standing Doe. A companion to *No. 2/8"/11 dec. of "Running Buck"* above. Sold separately or as a pair.

No. 2/8"/42 dec. Black Glass "Forget-me-not" Plate with hand painted white enamel scene of Girl with Sled. This decorated, triple "Forget-me-not" border plate is a companion to the *No. 2/8"/43 dec.* at right.

No. 2/8"/43 dec. Black Glass "Forget-me-not" Plate, hand painted white enamel scene of Boy Skating with Dog. *No. 2/8" "Forget-me-not"* sold singly or as a set with *No. 2/8"/42 "Girl with Sled"* at left.

29

Westmoreland Glass original company catalog 1973.

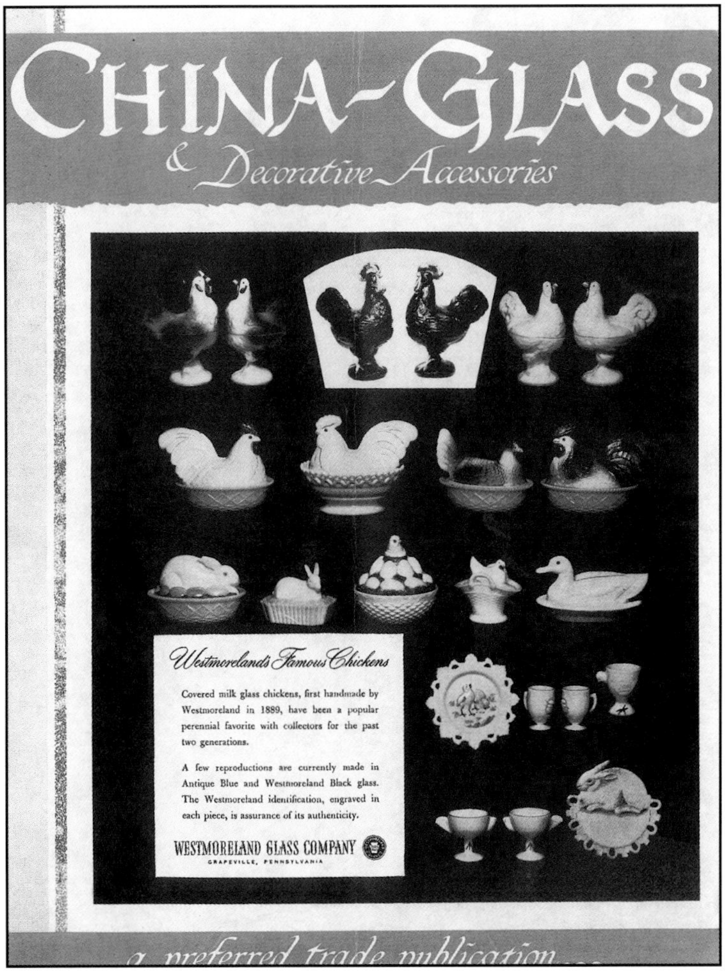

Westmoreland Glass black chickens and milk glass as featured on the front cover of *China, Glass and Decorative Accessories*, mid-1950s.

Westmoreland Glass advertisement appearing in *Crockery and Glass Journal*, February 1949.

Westmoreland Glass advertisement appearing in *Crockery and Glass Journal*, February 1930.

Westmoreland Glass original 1952 company catalog page.

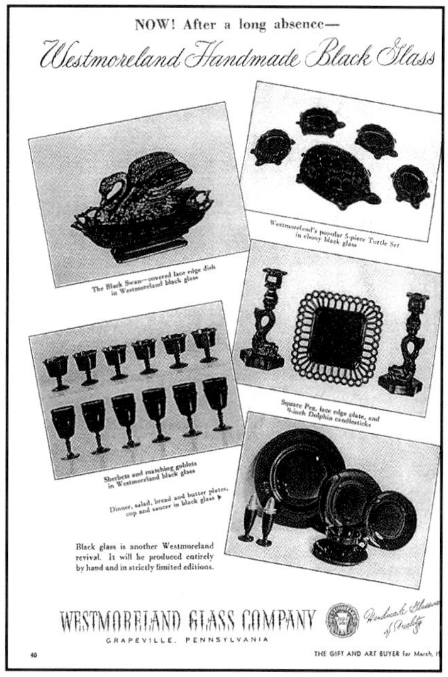

Westmoreland Glass advertisement appearing in *The Gift and Art Buyer*, March 1949.

Westmoreland Glass advertisement appearing in *The Pottery, Glass & Brass Salesman*, May 25, 1916.

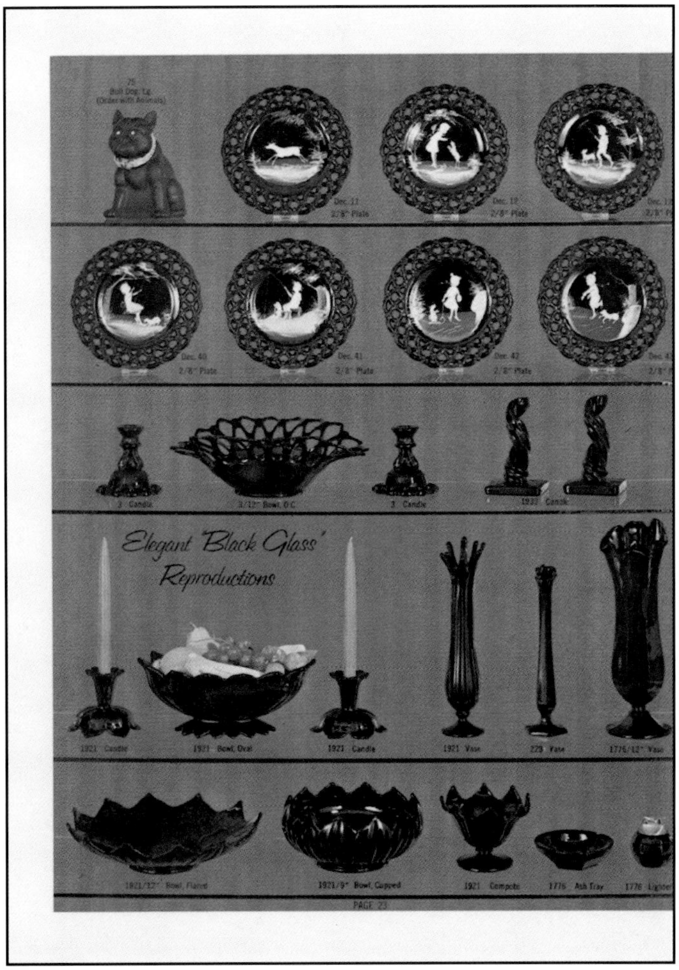

Westmoreland Glass original company catalog page circa 1972.

A page from the promotional brochure "The Remarkable Story of Handmade Glass by Westmoreland," 1970.

Westmoreland Glass advertisement appearing in *Crockery and Glass Journal*, February 24, 1916.

Westmoreland Glass Modernistic #700 line appearing in Montgomery Ward, Fall & Winter 1930-1931 catalog.

Chapter 15
Decorators & Distributors

When black glass attained one of its peaks in popularity in the late 1920s and into the 1930s a number of firms who decorated but did not manufacture glass entered the black glass market. Lotus Glass of Barnesville, Ohio, is a leading example of this type of black glass producer.

A second source of attribution are firms who distributed glass, predominately wholesalers but retailers as well. Several of these had black decorated glass made for them and then marketed it under their name.

Here we offer a modest selection of advertisements and catalog pages for some of those companies, both decorators and distributors. There is presently little information available about these "middle men" firms and we hope their inclusion here will generate a new interest and fuel research.

Advertisement for Edmondson Warrin, Inc. of NY and their black glass as offered in *The Pottery, Glass & Brass Salesman* of December 13, 1930.

Advertisement for Edmondson Warrin, Inc. of NY and their black glass as offered in *Crockery & Glass Journal* of January 1931.

One of the more recognized etched decorations on black glass is Black Forest made for Frank L. Van Deman & Son of New York City. The pieces here are all on glass blanks manufactured by Paden City Glass. *Crockery and Glass Journal*, March 1931.

Black Forest deep plate etched glassware manufactured for Frank L. Van Deman & Son of New York City as featured in the July 1930 issue of *The Crockery and Glass Journal*.

"The New Jet Black Decorative Glass Ware" marketed by Geo. Borgfeldt & Co. of New York, a wholesaler and not a decorator or manufacturer. Many of these shapes appear to be the products of U.S. Glass Co. *The Pottery, Glass & Brass Salesman*, March 23, 1916.

An exceptionally high quality silver decoration, promoted as "non tarnishing sterling silver," called Call of The Wild by Lotus Glass Co., decorators of Barnesville, Ohio. Original catalog page circa 1930s.

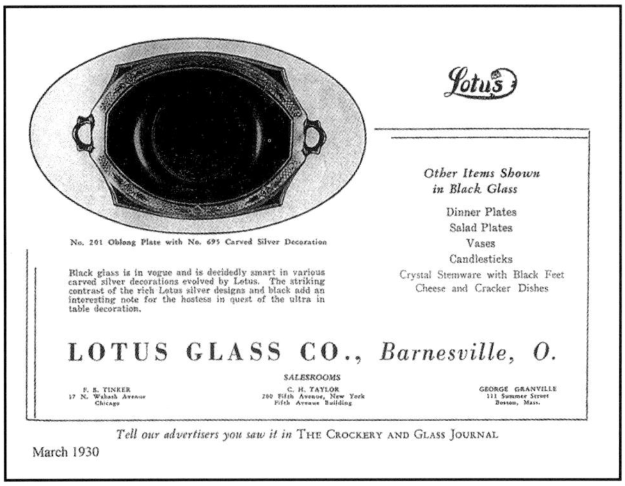

Advertisement for Lotus Glass Co., Silver Decoration on a Central Glass blank. *The Crockery and Glass Journal*, March 1930.

Left:
Advertisement announcing Lotus Glass Co.'s Non-Tarnishing sterling silver decoration. They claim to be "the first glassware decorator to offer Non-Tarnishing Sterling Silver decorations" by a new patented process. The decoration shown is Call of the Wild. *The Crockery and Glass Journal*, March 1931.

Right:
Advertisement for "the spirit of the age" – three bar ware pieces boldly offered for sale in the midst of America's prohibition, 1920-1934. Amazingly when these were being offered for sale it was illegal to own, transport or consume alcohol in the U.S. The middle bottle is a Morgantown Glass product and the pinch bottle probably Cambridge Glass. The floral/poppy? decoration is not commonly illustrated. *Crockery and Glass Journal*, November 1930.

An exceptionally high quality silver decoration, promoted as "non tarnishing sterling silver," called Call of The Wild by Lotus Glass Co., decorators of Barnesville, Ohio. Original catalog page circa 1930s. To learn more about this decoration, see the book *Lotus: Depression Glass and Far Beyond* listed in the bibliography.

An exceptionally high quality silver decoration, promoted as "non tarnishing sterling silver," called Call of The Wild by Lotus Glass Co., decorators of Barnesville, Ohio. Original catalog page circa 1930s.

A variety of Sterling Silver Deposit decorations on ebony blanks decorated by The Lotus Glass Co. of Barnesville, Ohio. Original catalog page circa 1930s.

Lola pattern Silver decoration on Ebony glass by Lotus Glass of Barnesville, Ohio. Original catalog page, circa 1930s.

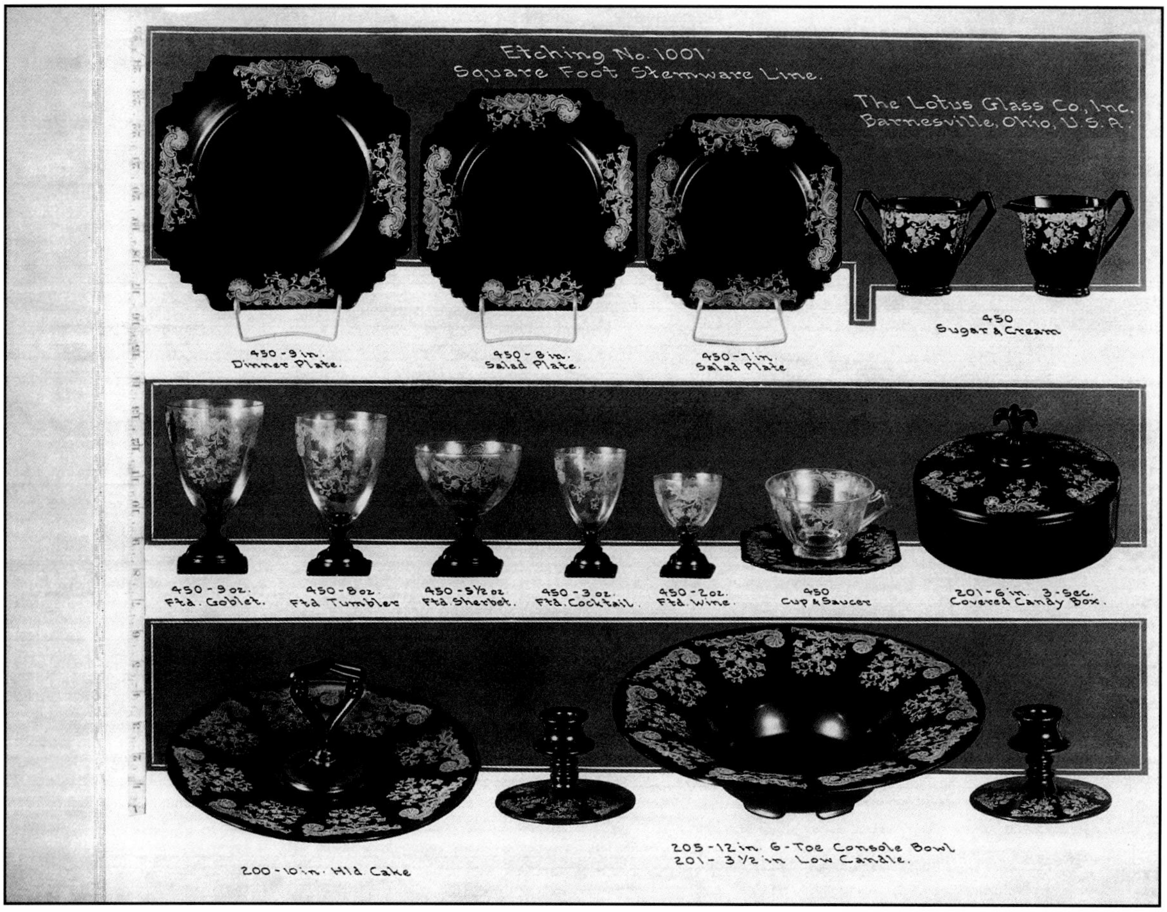

Etching 1001 named McGuire by glass author Hazel Marie Weatherman. The blanks are all Central Glass. Original catalog page circa 1930s.

A selection of hand painted Buffet Sets decorated by Lotus Glass Co. prior to dropping the word "cut" from its name in 1924. Original catalog page circa early 1920s. Note the sets are called "Buffet Sets" to go on the long buffets then in vogue in every dining room.

Wheeling Decorating Co. of Wheeling, West Virginia, offerings including heavy gold decoration on black glass. *Crockery & Glass Journal*, January 1931.

Black glass in an "all over encrusted gold decoration" as offered by Wheeling Decorating Co. of Wheeling, West Virginia. *Crockery and Glass Journal*, August 1932.

Diamond shaped candy box of ebony glass, shown at the left, decorated with gold or white gold. One of a group of unusually attractive items in the Corinthian Decorated Glass line, sold exclusively by R. D. Browd & Co., 1165 Broadway, New York.

Diamond shaped candy box offered with decoration of gold or white gold. Part of the "Corinthian Decorated Glass" line sold exclusively by R.D. Browd & Co." of New York. *Crockery and Glass Journal*, September 1930.

The West Virginia Museum of American Glass publishes the only nationwide magazine for glass collectors. To learn more about this publication, *All About Glass*, visit www.AllAboutGlass.org.

Bibliography

The following books were consulted in determining the attributions and other data presented in this book. Note that the parenthetical reference at the end of the usual bibliographic reference indicates how we have referenced each work in the text in this work.

Books

Baker, Larry and Bert Kennedy. *The Morgantown Etching Plates*. Mesquite, Texas: Bert Kennedy, 2003. (Baker & Kennedy 2003)

Cambridge Glass Company. *Cambridge Glass, 1927-1929*. Springfield, Ohio: Reprinted by Bill and Phyllis Smith, 1986. (Cambridge Catalog #27)

Coe, Debbie and Randy. *Glass Animals & Figurines*. Atglen, Pennsylvania: Schiffer Publishing, 2003. (Coe 2003)

Domitz, Carrie and Jerry. *Encyclopedia of Paden City Glass*. Paducah, Kentucky: Collector Books, 2004. (Domitz 2004)

Felt, Tom, Elaine and Rich Stoer. *The Glass Candlestick Book: Volume 1, Akro Agate to Fenton*. Paducah, Kentucky: Collector Books, 2003. (Felt and Stoer 2003)

Felt, Tom, Elaine, and Rich Stoer. *The Glass Candlestick Book: Volume 2, Fostoria to Jefferson*. Paducah, Kentucky: Collector Books, 2003. (Felt and Stoer 2003)

Felt, Tom, Elaine, and Rich Stoer. *The Glass Candlestick Book: Volume 3, Kanawha to Wright*. Paducah, Kentucky: Collector Books, 2005. (Felt and Stoer 2005)

Fry (H. C.) Glass Society. *The Collector's Encyclopedia of Fry Glassware*. 2nd ed. Paducah, Kentucky: Collector Books, 1999. (Fry Society 1999)

Gallagher, Jerry. *A Handbook of Old Morgantown Glass: Volume I, A Guide to Identification and Shape*. Morgantown, West Virginia: Old Morgantown Glass Collectors' Guild, 2001. (Gallagher 2001)

Garrison, Myrna and Bob. *Candlewick: Colored and Decorated Lines*. Atglen, Pennsylvania: Schiffer Publishing, 2003. (Garrison 2003)

Hardy, Roger and Claudia. *The Complete Line of the Akro Agate Company*. Clarksburg, West West Virginia: Roger Hardy, 1992. (Hardy 1992)

Heacock, William, James Measell and Berry Wiggins. *Dugan/Diamond: The Story of Indiana, Pennsylvania, Glass*. Marietta, Ohio: Antique Publications, 1993. (Heacock et al. 1993)

Kovar, Lorraine. *Westmoreland Glass: the Popular Years, 1940-1985*. Paducah, Kentucky: Collector Books, 2004. (Kovar 2004)

Long, Milbra and Emily Seate. *Fostoria Tableware, 1924-1943: The Crystal for America*. Paducah, Kentucky: Collector Books, 1999. (Long & Seate 1999)

Long, Milbra and Emily Seate. *Fostoria, Useful and Ornamental: The Crystal for America*. Paducah, Kentucky: Collector Books, 2000. (Long & Seate 2000)

Long, Milbra and Emily Seate. *The Fostoria Value Guide*. Paducah, Kentucky: Collector Books, 2003. (Long & Seate 2003)

Mauzy, Barbara and Jim. *Mauzy's Kitchen Collectibles*. Atglen, Pennsylvania: Schiffer Publishing, 2004. (Mauzy 2004)

Measell, James. *New Martinsville Glass, 1900-1944*. Marietta, Ohio: Antique Publications, 1994. (Measell 1994)

Measell, James and Berry Wiggins. *Great American Glass of the Roaring 20s & Depression Era*. Marietta, Ohio: Antique Publications, 1998. (Measell and Wiggins 1998)

Murschell, Dale L. *Glass of Cumberland Maryland Area*. Springfield, West Virginia: Dale Murschell, 2002. (Murschell 2002)

National Cambridge Collectors, Inc. *The Cambridge Glass Company: [catalog reprints]*. Cambridge, Ohio: National Cambridge Collectors, 1976. (Cambridge Collectors 1976)

Page, Bob and Dale Frederiksen. *Crystal Stemware Identification Guide*. Paducah, Kentucky: Collector Books, 1997. (Page & Frederiksen 1997)

Page, Bob and Dale Frederiksen. *Seneca Glass Company, 1891-1983: A Stemware Identification Guide*. Greensboro, North Carolina: Page-Frederiksen Publishing Company, 1995. (Page & Frederiksen 1995)

Page, Bob and Dale Frederiksen. *Tiffin Is Forever: A Stemware Identification Guide*. Greensboro, North Carolina: Page-Frederiksen Publishing Company, 1994. (Page & Frederiksen 1994)

Schmidt, Tim. *Central Glass Works: The Depression era*. Atglen, Pennsylvania: Schiffer Publishing, 2004. (Schmidt 2004)

Six, Dean. *Lotus: Depression Glass & Far Beyond*. Atglen, Pennsylvania: Schiffer Publishing, 2005. (Dean 2005)

Six, Dean. *Viking Glass, 1944-1970*. Atglen, Pennsylvania: Schiffer Publishing, 2003. (Six 2003)

Six, Dean, *West Virginia Glass Between the World Wars*. Atglen, Pennsylvania: Schiffer Publishing, 2002. (Six 2002)

Snyder, Jeffrey B. *Morgantown Glass: From Depression Glass Through the 1960s*. Atglen, Pennsylvania: Schiffer Publishing,1998. (Snyder 1998)

Spencer, Dick and Pat. *Glass Animals Including Animal & Figural Related Items*. 2nd ed. Paducah, Kentucky: Collector Books, 2004. (Spencer 2004)

Stout, Sandra McPhee. *The Complete Book of McKee Glass*. North Kansas City, Kansas: Trojan Press, 1972. (McKee 1972)

Teal, Ron, Sr. *Tiara Exclusives: Company Catalog Reprints*. Marietta, Ohio: Antique Publications, 2000. (Teal 2000)

Toohey, Marlena. *A Collector's Guide to Black Glass 2*. Marietta, Ohio: Antique Publications, 1999. (Toohey 1999)

Walk, John. *Fenton Glass Compendium, 1970-1985*. Atglen, Pennsylvania: Schiffer Publishing, 2001. (Walk 2001)

Weatherman, Hazel Marie. *Colored Glassware of the Depression Era 2*. Springfield, Missouri: Weatherman Glassbooks, 1974. (Weatherman 1974)

Whitmyer, Margaret and Kenn. *Fenton Art Glass, 1907-1939: Identification & Value Guide*. Paducah, Kentucky: Collector Books, 1996. (Whitmyer 1996)

Wilson, Jack D. *Phoenix & Consolidated Art Glass, 1926-1980*. Marietta, Ohio: Antique Publications, 1989. (Wilson 1989)

Websites

Kappenman, Beverly. *Black Glass*. http://members.aol.com/blkapp

Replacements, Ltd. http://www.replacements.com

Trade Journals

China, Glass and Decorative Accessories

China, Glass and Lamps

Crockery and Glass Journal

Pottery, Glass & Brass Salesman

Index